Dairy Science and Technology

Dairy Science and Technology

Edited by
Drake Ward

Larsen & Keller
www.larsen-keller.com

Dairy Science and Technology
Edited by Drake Ward
ISBN: 978-1-63549-083-1 (Hardback)

☰ Larsen & Keller

Published by Larsen and Keller Education,
5 Penn Plaza,
19th Floor,
New York, NY 10001, USA

Cataloging-in-Publication Data

Dairy science and technology / edited by Drake Ward.
    p. cm.
Includes bibliographical references and index.
ISBN 978-1-63549-083-1
1. Dairying. 2. Dairy processing. 3. Dairy products. 4. Dairy farming.
5. Dairy engineering. I. Ward, Drake.
SF239 .D35 2017
637--dc23

The publisher's policy is to use permanent paper from mills that operate a sustainable forestry policy. Furthermore, the publisher ensures that the text paper and cover boards used have met acceptable environmental accreditation standards.

Printed and bound in the United States of America.

For more information regarding Larsen and Keller Education and its products, please visit the publisher's website www.larsen-keller.com

Table of Contents

Preface

This book explores all the important aspects of dairy science in the present day scenario. It discusses in detail the concepts and practical applications of this subject. Dairy science refers to the study of milk production along with the production of cheese, yogurt, butter, etc. A dairy science professional also studies the operations and management of a cattle ranch and a dairy plant. The text is a valuable compilation of topics, ranging from the basic to the most complex theories and principles in the field of dairy science and technology. While understanding the long-term perspectives of the topics, the book makes an effort in highlighting their impact as a modern tool for the growth of the discipline. It will provide comprehensive knowledge to the students. Through this textbook, we attempt to further enlighten the readers about the new concepts in this field.

Given below is the chapter wise description of the book:

Chapter 1- Dairy farming is the production of milk for consumption as well as commercial purposes. The animals used in commercial dairy farming are goats, camels, cows and sheeps. Dairy farming has been a part of agriculture over a period of thousand years. The chapter on dairy farming offers an insightful focus, keeping in mind the subject matter.

Chapter 2- Cream is a product that is produced by the dairy. Cream is also dried and then made into a powder and sold into the market. The primary dairy products that are used on a daily basis are sour cream, crème fraiche, butter and churning. The major categories of cream and butter are dealt with great details in the following chapter.

Chapter 3- Milk is the primary source of nutrition; as an agricultural product it is extracted from animals such as cows, goats and camels. Homogenization, pasteurization, fouling, soured milk, raw milk and automatic milking are some of the aspects of milk and milk processing. The topics discussed in the section are of great importance to broaden the existing knowledge on milk.

Chapter 4- Powdered milk is a dairy product that is dried milk. One of the main purposes of drying milk is by preserving it. Some of the methods are spray drying, drum drying and freeze-drying. The aspects elucidated in this chapter are of vital importance and provides a better understanding of powdered milk manufacturing.

Chapter 5- Yogurt is a food product that is produced from milk. The other dairy products discussed are strained yogurt, doogh, curd and buffalo curd. The topics discussed in the section help the readers in broadening the existing knowledge on yogurt and curd.

Chapter 6- Cheese is a food item that is derived from milk and is produced in different flavors. There are different types of cheese; almost 500 different varieties of cheese are produced. The process used in making cheese is known as cheese ripening. This section will provide an integrated understanding of cheese and cheese making processes.

Chapter 7- Fermented milk products are the products that are fermented by using lactic acid bacteria such as lactobacillus and leuconostoc. The production of milk products can be traced back to around 10,000 BC. The products discussed in this section are filmjölk, kumis, kefir, chal, lassi and yakult. The main products of dairy products are discussed in the following chapter.

Indeed, my job was extremely crucial and challenging as I had to ensure that every chapter is informative and structured in a student-friendly manner. I am thankful for the support provided by my family and colleagues during the completion of this book.

Editor

Understanding Dairy Farming

Dairy farming is the production of milk for consumption as well as commercial purposes. The animals used in commercial dairy farming are goats, camels, cows and sheeps. Dairy farming has been a part of agriculture over a period of thousand years. The chapter on dairy farming offers an insightful focus, keeping in mind the subject matter.

Dairy Farming

A rotary milking parlor at a modern dairy facility, located in Germany

Dairy farming is a class of agriculture for long-term production of milk, which is processed (either on the farm or at a dairy plant, either of which may be called a dairy) for eventual sale of a dairy product.

Common Species

Although any mammal can produce milk, commercial dairy farms are typically one-species enterprises. In developed countries, dairy farms typically consist of high producing dairy cows. Other species used in commercial dairy farming include goats, sheep, and camels. In Italy, donkey dairies are growing in popularity to produce an alternative milk source for human infants.

A dairy farm on the banks of the Columbia River in Clark County, Washington (May 1973).

History

Dairy farming has been part of agriculture for thousands of years. Historically it has been one part of small, diverse farms. In the last century or so larger farms doing only dairy production have emerged. Large scale dairy farming is only viable where either a large amount of milk is required for production of more durable dairy products such as cheese, butter, etc. or there is a substantial market of people with cash to buy milk, but no cows of their own.

Hand milking

Woman hand milking a cow.

Centralized dairy farming as we understand it primarily developed around villages and cities, where residents were unable to have cows of their own due to a lack of grazing land. Near the town, farmers could make some extra money on the side by having additional animals and selling the milk in town. The dairy farmers would fill barrels with milk in the morning and bring it to market on a wagon. Until the late 19th century, the milking of the cow was done by hand. In the United States, several large dairy operations existed in some northeastern states and in the west, that involved as many as several hundred cows, but an individual milker could not be expected to milk more than a dozen cows a day. Smaller operations predominated.

For most herds, milking took place indoors twice a day, in a barn with the cattle tied by the neck with ropes or held in place by stanchions. Feeding could occur simultaneously with milking in the barn, although most dairy cattle were pastured during the day between milkings. Such examples of this method of dairy farming are difficult to locate, but some are preserved as a historic site for a glimpse into the days gone by. One such instance that is open for this is at Point Reyes National Seashore.

Dairy farming has been part of agriculture for thousands of years. Historically it has been one part of small, diverse farms. In the last century or so larger farms doing only dairy production have emerged. Large scale dairy farming is only viable where either a large amount of milk is required for production of more durable dairy products such as cheese, butter, etc. or there is a substantial market of people with cash to buy milk, but no cows of their own.

Vacuum Bucket Milking

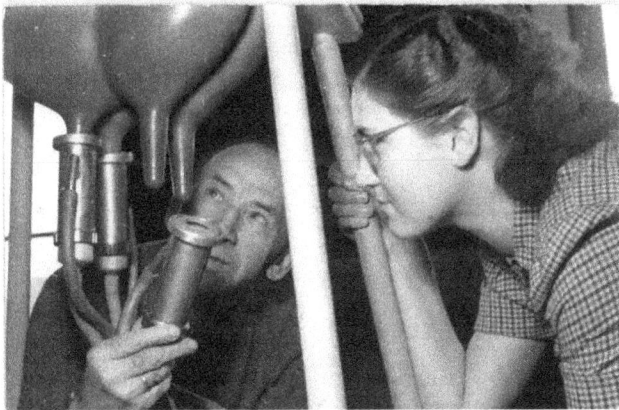

Demonstration of a new Soviet milker device. East Germany, 1952

The first milking machines were an extension of the traditional milking pail. The early milker device fit on top of a regular milk pail and sat on the floor under the cow. Following each cow being milked, the bucket would be dumped into a holding tank. These were introduced in the early 20th century.

This developed into the Surge hanging milker. Prior to milking a cow, a large wide leather strap called a surcingle was put around the cow, across the cow's lower back. The milker device and collection tank hung underneath the cow from the strap. This innovation allowed the cow to move around naturally during the milking process rather than having to stand perfectly still over a bucket on the floor.

Milking Pipeline

The next innovation in automatic milking was the milk pipeline, introduced in the late 20th century. This uses a permanent milk-return pipe and a second vacuum pipe that encircles the barn or milking parlor above the rows of cows, with quick-seal entry ports above each cow. By eliminating the need for the milk container, the milking device shrank in size and weight to the point where it could hang under the cow, held up only by the sucking force of the milker nipples on the cow's udder. The milk is pulled up into the milk-return pipe by the vacuum system, and then flows by gravity to the milkhouse vacuum-breaker that puts the milk in the storage tank. The pipeline system greatly reduced the physical labor of milking since the farmer no longer needed to carry around huge heavy buckets of milk from each cow.

The pipeline allowed barn length to keep increasing and expanding, but after a point farmers started to milk the cows in large groups, filling the barn with one-half to one-third of the herd, milking the animals, and then emptying and refilling the barn. As herd sizes continued to increase, this evolved into the more efficient milking parlor.

Milking Parlors

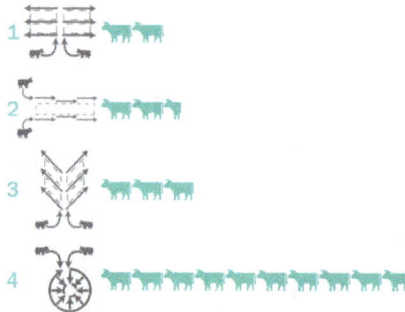

Efficiency of four different milking parlors.
1=Bali-Style 50 cows/h.   2=Swingover 60 cows/h.   3=Herringbone 75 cows/h.   4=Rotary 250 cows/h.

Innovation in milking focused on mechanizing the milking parlor (known in Australia and New Zealand as a *milking shed*) to maximize the number of cows per operator which streamlined the milking process to permit cows to be milked as if on an assembly line, and to reduce physical stresses on the farmer by putting the cows on a platform slightly above the person milking the cows to eliminate having to constantly bend over. Many older and smaller farms still have tie-stall or stanchion barns, but worldwide a majority of commercial farms have parlors.

Herringbone and Parallel Parlors

In herringbone and parallel parlors, the milker generally milks one row at a time. The milker will move a row of cows from the holding yard into the milking parlor, and milk each cow in that row. Once all of the milking machines have been removed from the milked row, the milker releases the cows to their feed. A new group of cows is then loaded into the now vacant side and the process repeats until all cows are milked. Depending on the size of the milking parlor, which normally is the bottleneck, these rows of cows can range from four to sixty at a time. The benefits of a herringbone parlour are easy maintenance, the durability, stability, and improved safety for animals and humans when compared to tie stall

Rotary Parlors

Rotary milking parlor

In rotary parlors, the cows are loaded one at a time onto the platform as it rotates. The milker stands near the entry to the parlor and puts the cups on the cows as they move past. By the time the platform has completed almost a full rotation, another milker or a machine removes the cups and the cow steps backwards off the platform and then walks to its feed. Rotary cowsheds, as they are called in New Zealand, started in the 1980s but are expensive compared to Herringbone cowshed - the older New Zealand norm.

Automatic Milker Take-off

It can be harmful to an animal for it to be over-milked past the point where the udder has stopped releasing milk. Consequently, the milking process involves not just applying the milker, but also monitoring the process to determine when the animal has been *milked out* and the milker should be removed. While parlor operations allowed a farmer to milk many more animals much more quickly, it also increased the number of animals to be monitored simultaneously by the farmer. The automatic take-off system was developed to remove the milker from the cow when the milk flow reaches a preset level, relieving the farmer of the duties of carefully watching over 20 or more animals being milked at the same time.

Fully Automated Robotic Milking

In the 1980s and 1990s, robotic milking systems were developed and introduced (principally in the EU). Thousands of these systems are now in routine operation. In these systems the cow has a high degree of autonomy to choose her time of milking freely during the day (some alternatives may apply, depending on cow-traffic solution used at a farm level). These systems are generally limited to intensively managed systems although research continues to match them to the requirements of grazing cattle and to develop sensors to detect animal health and fertility automatically. Every time the cow enters the milking unit she is fed concentrates and her collar is scanned to record production data.

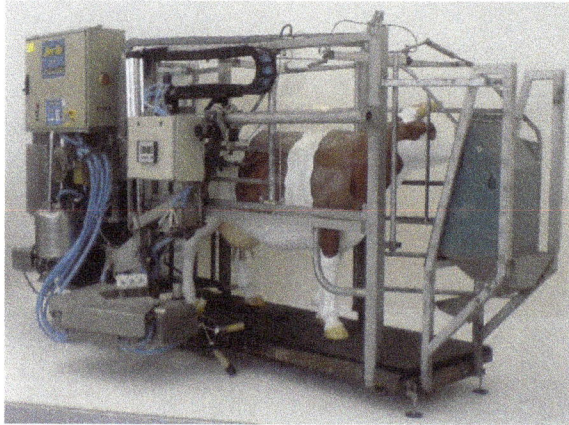

An automatic milking system unit as an exhibit at a museum

History of Milk Preservation Methods

Cool temperature has been the main method by which milk freshness has been extended. When windmills and well pumps were invented, one of their first uses on the farm, besides providing water for animals themselves, was for cooling milk, to extend its storage life, until it would be transported to the town market.

The naturally cold underground water would be continuously pumped into a cooling tub or vat. Tall, ten-gallon metal containers filled with freshly obtained milk, which is naturally warm, were placed in this cooling bath. This method of milk cooling was popular before the arrival of electricity and refrigeration.

Refrigeration

When refrigeration first arrived (the 19th century) the equipment was initially used to cool cans of milk, which were filled by hand milking. These cans were placed into a cooled water bath to remove heat and keep them cool until they were able to be transported to a collection facility. As more automated methods were developed for harvesting milk, hand milking was replaced and, as a result, the milk can was replaced by a bulk milk cooler. 'Ice banks' were the first type of bulk milk cooler. This was a double wall vessel with evaporator coils and water located between the walls at the bottom and sides of the tank. A small refrigeration compressor was used to remove heat from the evaporator coils. Ice eventually builds up around the coils, until it reaches a thickness of about three inches surrounding each pipe, and the cooling system shuts off. When the milking operation starts, only the milk agitator and the water circulation pump, which flows water across the ice and the steel walls of the tank, are needed to reduce the incoming milk to a temperature below 5 degrees.

This cooling method worked well for smaller dairies, however was fairly inefficient and was unable to meet the increasingly higher cooling demand of larger milking parlors. In the mid-1950s direct expansion refrigeration was first applied directly to the bulk milk cooler. This type of cooling utilizes an evaporator built directly into the inner wall of the storage tank to remove heat from the milk. Direct expansion is able to cool milk at a much faster rate than early ice bank type coolers and is still the primary method for bulk tank cooling today on small to medium-sized operations.

Another device which has contributed significantly to milk quality is the plate heat exchanger (PHE). This device utilizes a number of specially designed stainless steel plates with small spaces between them. Milk is passed between every other set of plates with water being passed between the balance of the plates to remove heat from the milk. This method of cooling can remove large amounts of heat from the milk in a very short time, thus drastically slowing bacteria growth and thereby improving milk quality. Ground water is the most common source of cooling medium for this device. Dairy cows consume approximately 3 gallons of water for every gallon of milk production and prefer to drink slightly warm water as opposed to cold ground water. For this reason, PHE's can result in drastically improved milk quality, reduced operating costs for the dairymen by reducing the refrigeration load on his bulk milk cooler, and increased milk production by supplying the cows with a source of fresh warm water.

Plate heat exchangers have also evolved as a result of the increase of dairy farm herd sizes in the United States. As a dairyman increases the size of his herd, he must also increase the capacity of his milking parlor in order to harvest the additional milk. This increase in parlor sizes has resulted in tremendous increases in milk throughput and cooling demand. Today's larger farms produce milk at a rate which direct expansion refrigeration systems on bulk milk coolers cannot cool in a timely manner. PHE's are typically utilized in this instance to rapidly cool the milk to the desired temperature (or close to it) before it reaches the bulk milk tank. Typically, ground water is still utilized to provide some initial cooling to bring the milk to between 55 and 70 °F (21 °C). A second (and sometimes third) section of the PHE is added to remove the remaining heat with a mixture of chilled pure water and propylene glycol. These chiller systems can be made to incorporate large evaporator surface areas and high chilled water flow rates to cool high flow rates of milk.

Milking Operation

Milking machines are held in place automatically by a vacuum system that draws the ambient air pressure down from 15 to 21 pounds per square inch (100 to 140 kPa) of vacuum. The vacuum is also used to lift milk vertically through small diameter hoses, into the receiving can. A milk lift pump draws the milk from the receiving can through large diameter stainless steel piping, through the plate cooler, then into a refrigerated bulk tank.

Milk is extracted from the cow's udder by flexible rubber sheaths known as liners or inflations that are surrounded by a rigid air chamber. A pulsating flow of ambient air and vacuum is applied to the inflation's air chamber during the milking process. When ambient air is allowed to enter the chamber, the vacuum inside the inflation causes the inflation to collapse around the cow's teat, squeezing the milk out of teat in a similar fashion as a baby calf's mouth massaging the teat. When the vacuum is reapplied in the chamber the flexible rubber inflation relaxes and opens up, preparing for the next squeezing cycle.

It takes the average cow three to five minutes to give her milk. Some cows are faster or slower. Slow-milking cows may take up to fifteen minutes to let down all their milk. Though milking speed is not related to the quality of milk produced by the cow, it does impact the management of the milking process. Because most milkers milk cattle in groups, the milker can only process a group of cows at the speed of the slowest-milking cow. For this reason, many farmers will group slow-milking cows so as not to stress the faster milking cows.

The extracted milk passes through a strainer and plate heat exchangers before entering the tank, where it can be stored safely for a few days at approximately 40 °F (4 °C). At pre-arranged times, a milk truck arrives and pumps the milk from the tank for transport to a dairy factory where it will be pasteurized and processed into many products. The frequency of pick up depends and the production and storage capacity of the dairy; large dairies will have milk pick-ups once per day.

Management of the Herd

Modern dairy farmers use milking machines and sophisticated plumbing systems to harvest and store the milk from the cows, which are usually milked two or three times daily. In New Zealand, some farmers seeking a better life style, are milking only once per day, trading a slight reduction in production of milk for increased leisure time. During the summer months, cows may be turned out to graze in pastures, both day and night, and are brought into the barn to be milked.

Barns may also incorporate tunnel ventilation into the architecture of the barn structure. This ventilation system is highly efficient and involves opening both ends of the structure allowing cool air to blow through the building. Farmers with this type of structure keep cows inside during the summer months to prevent heat stress, sunburn and damage to udders. During the winter months the cows may be kept in the barn, which is warmed by their collective body heat. Even in winter, the heat produced by the cattle requires the barns to be ventilated for cooling purposes. Many large, modern facilities, and particularly those in tropical areas, keep all animals inside at all times to facilitate herd management.

Farmers typically grow their own food for their cattle. Crops grown may include corn, alfalfa, timothy, wheat, oats, sorghum and clover. These plants are often processed after harvest to preserve or improve nutrient value and prevent spoiling. Corn, alfalfa, wheat, oats, and sorghum crops are often anaerobically fermented to create silage. Many crops such as alfalfa, timothy, oats, and clover are allowed to dry in the field after cutting before being baled into hay.

In the southern hemisphere such as in Australia and New Zealand, cows spend most of their lives outside on pasture, although they may receive supplementation during periods of low pasture availability. Typical supplementary feeds in Australasia are hay, silage or ground maize. The trend in New Zealand is towards feeding cows on a concrete pad to prevent loss of feed by trampling. In New Zealand slower growing winter pasture is rationed. It is carefully controlled by light weight portable electric break feeding fences run on mains power that can be easily repositioned.

Concerns

Animal Waste from Large Cattle Dairies

Dairy CAFO—EPA

As measured in phosphorus, the waste output of 5,000 cows roughly equals a municipality of 70,000 people. In the U.S., dairy operations with more than 1,000 cows meet the EPA definition of a CAFO (Concentrated Animal Feeding Operation), and are subject to EPA regulations. For example, in the San Joaquin Valley of California a number of dairies have been established on a very large scale. Each dairy consists of several modern milking parlor set-ups operated as a single enterprise. Each milking parlor is surrounded by a set of 3 or 4 loafing barns housing 1,500 or 2,000 cattle. Some of the larger dairies have planned 10 or more series of loafing barns and milking parlors in this arrangement, so that the total operation may include as many as 15,000 or 20,000 cows. The milking process for these dairies is similar to a smaller dairy with a single milking parlor but repeated several times. The size and concentration of cattle creates major environmental issues associated with manure handling and disposal, which requires substantial areas of cropland (a ratio of 5 or 6 cows to the acre, or several thousand acres for dairies of this size) for manure spreading and dispersion, or several-acre methane digesters. Air pollution from methane gas associated with manure management also is a major concern. As a result, proposals to develop dairies of this size can be controversial and provoke substantial opposition from environmentalists including the Sierra Club and local activists.

The potential impact of large dairies was demonstrated when a massive manure spill occurred on a 5,000-cow dairy in Upstate New York, contaminating a 20-mile (32 km) stretch of the Black River, and killing 375,000 fish. On 10 August 2005, a manure storage lagoon collapsed releasing 3,000,000 US gallons (11,000,000 l; 2,500,000 imp gal) of manure into the Black River. Subsequently the New York Department of Environmental Conservation mandated a settlement package of $2.2 million against the dairy.

When properly managed, dairy and other livestock waste, due to its nutrient content (N, P, K), makes an excellent fertilizer promoting crop growth, increasing soil organic matter, and improving overall soil fertility and tilth characteristics. Most dairy farms in the United States are required to develop nutrient management plans for their farms, to help balance the flow of nutrients and reduce the risks of environmental pollution. These plans encourage producers to monitor all nutrients coming onto the farm as feed, forage, animals, fertilizer, etc. and all nutrients exiting the farm as product, crop, animals, manure, etc. For example, a precision approach to animal feeding results in less overfeeding of nutrients and a subsequent decrease in environmental excretion of nutrients, such as phosphorus. In recent years, nutritionists have realized that requirements for phosphorus are much lower than previously thought. These changes have allowed dairy producers to reduce the amount of phosphorus being fed to their cows with a reduction in environmental pollution.

Use of Hormones

It is possible to maintain higher milk production by supplementing cows with growth hormones known as recombinant BST or rBST, but this is controversial due to its effects on animal and possibly human health. The European Union, Japan, Australia, New Zealand and Canada have banned its use due to these concerns.

In the US however, no such prohibition exists, and approximately 17.2% of dairy cows are treated in this way. The U.S. Food and Drug Administration states that no "significant difference" has been found between milk from treated and non-treated cows but based on consumer concerns several milk purchasers and resellers have elected not to purchase milk produced with rBST.

Animal Welfare

The practice of dairy production in a factory farm environment has been criticized by animal welfare activists. Some of the ethical complaints regarding dairy production cited include how often the dairy cattle must remain pregnant, the separation of calves from their mothers, how dairy cattle are housed and environmental concerns regarding dairy production.

The production of milk requires that the cow be in lactation, which is a result of the cow having given birth to a calf. The cycle of insemination, pregnancy, parturition, and lactation, followed by a "dry" period of about two months of forty-five to fifty days, before calving which allows udder tissue to regenerate. A dry period that falls outside this time frame can result in decreased milk production in subsequent lactation.

An important part of the dairy industry is the removal of the calves off the mother's milk after the three days of needed colostrum, allowing for the collection of the milk produced. On some dairies, in order for this to take place, the calves are fed milk replacer, a substitute for the whole milk produced by the cow. Milk replacer is generally a powder, which comes in large bags, and is added to precise amounts of water, and then fed to the calf via bucket or bottle. However, not all dairies use milk replacer - some continue to feed calves milk from the cows in the milking herd. Some dairies even pasteurize extra milk from the main herd to feed calves.

Milk replacers are classified by three categories: protein source, protein/fat (energy) levels, and medication or additives (e.g. vitamins and minerals). Proteins for the milk replacer come from different sources; the more favorable and more expensive all milk protein (e.g. whey protein- a by-product of the cheese industry) and alternative proteins including soy, animal plasma and wheat gluten. The ideal levels for fat and protein in milk replacer are 10-28% and 18-30%, respectively. The higher the energy levels (fat and protein), the less starter feed (feed which is given to young animals) the animal will consume. Weaning can take place when a calf is consuming at least two pounds of starter feed a day and has been on starter for at least three weeks. Milk replacer has climbed in cost US$15–20 a bag in recent years, so early weaning is economically crucial to effective calf management.

Because of the danger of infection to humans, it is important to maintain the health of milk-producing cattle. Common ailments affecting dairy cows include infectious disease (e.g. mastitis, endometritis and digital dermatitis), metabolic disease (e.g. milk fever and ketosis) and injuries caused by their environment (e.g. hoof and hock lesions).

Lameness is commonly considered one of the most significant animal welfare issues for dairy cattle, and is best defined as any abnormality that causes an animal to change its gait. It can be caused by a number of sources, including infections of the hoof tissue (e.g. fungal infections that cause dermatitis) and physical damage causing bruising or lesions (e.g. ulcers or hemorrhage of the hoof). Housing and management features common in modern dairy farms (such as concrete barn floors, limited access to pasture and suboptimal bed-stall design) have been identified as contributing risk factors to infections and injuries. New dairy farms being built now include non-slip flooring and other features designed to minimize risk to cows when moving between pens and to the milking parlor.

Market

Worldwide

Holstein cows on a dairy farm, Comboyne, New South Wales

Dairy farm in Võru Parish, Estonia

There is a great deal of variation in the pattern of dairy production worldwide. Many countries which are large producers consume most of this internally, while others (in particular New Zealand), export a large percentage of their production. Internal consumption is often in the form of liquid milk, while the bulk of international trade is in processed dairy products such as milk powder.

The milking of cows was traditionally a labor-intensive operation and still is in less developed countries. Small farms need several people to milk and care for only a few dozen cows, though for many farms these employees have traditionally been the children of the farm family, giving rise to the term "family farm".

Advances in technology have mostly led to the radical redefinition of "family farms" in industrialized countries such as Australia, New Zealand, and the United States. With farms of hundreds of cows producing large volumes of milk, the larger and more efficient dairy farms are more able to weather severe changes in milk price and operate profitably, while "traditional" family farms generally do not have the equity or income other larger scale farms do. The common public perception of large corporate farms supplanting smaller ones is generally a misconception, as many small family farms expand to take advantage of economies of scale, and incorporate the business to limit the legal liabilities of the owners and simplify such things as tax management.

Before large scale mechanization arrived in the 1950s, keeping a dozen milk cows for the sale of milk was profitable. Now most dairies must have more than one hundred cows being milked at a time in order to be profitable, with other cows and heifers waiting to be "freshened" to join the milking herd . In New Zealand the average herd size, for the 2009/2010 season, is 376 cows.

Worldwide, the largest milk producer is India (more than 55% buffalo milk), the largest cow milk exporter is New Zealand, and the largest importer is China. The European Union with its present 28 member countries produced 158,800,000 metric tons (156,300,000 long tons; 175,000,000 short tons) in 2013(96.8% cow milk), the most by any politico-economic union.

Supply Management

In Canada the dairy industry a system was put in by the Canadian Dairy Commission that is responsible for maintaining a stable market and line of employment for farmers by using a Supply Management System. This system was put into play in the early 1970s for a consistent pricing of milk for farmers, ensuring no fluctuation in the market The prices are based on the demand for milk throughout the country and how much is being produced. The Canadian Milk Supply Management Committee is in charge of monitoring the production rates of milk. In order to start a new farm or increase production more share into the SMS needs to be bought into known as "Quota". Quota is a fixed amount that an individual or group are bound to produce or receive. in this case farmers must remain up to or below the amount of "quota" they have bought share of. Each province in Canada has their own cap on quota based on the demand in the market There is a cap on the countries quota known as total quota per month. In the month of 2016 the total kgs of butter fat produced per month was 28,395,848

World total milk production in 2009 FAO statistics (including cow/buffalo/goat/sheep/camel milk)		
Rank	**Country**	**Production (10^6 kg/y)**
	World	**696,554**
1	India	110,040
2	United States	85,859
3	China	40,553
4	Pakistan	34,362
5	Russia	32,562
6	Germany	28,691

7	Brazil	27,716
8	France	24,218
9	New Zealand	15,217
10	United Kingdom	13,237
11	Italy	12,836
12	Turkey	12,542
13	Poland	12,467
14	Ukraine	11,610
15	Netherlands	11,469
16	Mexico	10,931
17	Argentina	10,500
18	Australia	9,388
19	Canada	8,213
20	Japan	7,909

European Union

Production building at a dairy farm in Norway.

The European Union with its present 27 member countries is the largest milk producer in the world. The largest producers within the EU are Germany and France.

Dairy production in the EU is heavily distorted due to the Common Agricultural Policy – being subsidized in some areas, and subject to production quotas in other.

European total milk production in 2009 FAO statistics (including cow/goat/sheep/buffalo milk)		
Rank	**Country**	**Production (10^6 kg/y)**
	European Union (all 27 countries)	**153,033**

1	Germany	28,691
2	France	24,218
3	United Kingdom	13,237
4	Italy	12,836
5	Poland	12,467
6	Netherlands	11,469
7	Spain	7,252
8	Romania	5,809
9	Ireland	5.373
10	Denmark	4,814

Israel

The dairy farm on Sa'ad was the Israeli leader in 2011 for productivity with an average of 13,785 litres (3,032 imp gal; 3,642 US gal) per head that year. A dairy cow named Kharta, was the world record holder giving 18,208 litres (4,005 imp gal; 4,810 US gal) liters of milk. The 954 Israeli dairy farms achieved a world leading average production of 11,775 litres (2,590 imp gal; 3,111 US gal) a year per head, while the national average per head was 10,336 litres (2,274 imp gal; 2,730 US gal). Israeli consumption is lower than other western countries with an average of 180 litres (40 imp gal; 48 US gal) per person.

United States

In the United States, the top five dairy states are, in order by total milk production; California, Wisconsin, New York, Idaho, and Pennsylvania. Dairy farming is also an important industry in Florida, Minnesota, Ohio and Vermont. There are 65,000 dairy farms in the United States.

Pennsylvania has 8,500 farms with 555,000 dairy cows. Milk produced in Pennsylvania yields an annual revenue of about US$1.5 billion.

Milk prices collapsed in 2009. Senator Bernie Sanders accused Dean Foods of controlling 40% of the country's milk market. He has requested the United States Department of Justice to pursue an anti-trust investigation. Dean Foods says it buys 15% of the country's raw milk. In 2011, a federal judge approved a settlement of $30 million to 9,000 farmers in the Northeast.

Herd size in the US varies between 1,200 on the West Coast and Southwest, where large farms are commonplace, to roughly 50 in the Midwest and Northeast, where land-base is a significant limiting factor to herd size. The average herd size in the U.S. is about one hundred cows per farm but the median size is 900 cows with 49% of all cows residing on farms of 1000 or more cows.

Dairy Cattle

A Holstein cow with prominent udder and less muscle than is typical of beef breeds

Dairy cattle (also called dairy cows or milk cows) are cattle cows bred for the ability to produce large quantities of milk, from which dairy products are made. Dairy cows generally are of the species *Bos taurus*.

Historically, there was little distinction between dairy cattle and beef cattle, with the same stock often being used for both meat and milk production. Today, the bovine industry is more specialized and most dairy cattle have been bred to produce large volumes of milk.

The United States dairy herd produced 83.9 billion kg (185 billion lbs) of milk in 2007, up from 52.6 billion kg (116 billion lbs) in 1950, yet there were only about 9 million cows on U.S. dairy farms—about 13 million fewer than there were in 1950. The top breed of dairy cow within Canada's national herd category is Holstein, taking up 93% of the dairy cow population, have an annual production rate of 10 257 kg of milk per cow that contains 3.9% butter fat and 3.2% protein.

Management

Dairy cows may be found either in herds or dairy farms where dairy farmers own, manage, care for, and collect milk from them, or on commercial farms. Herd sizes vary around the world depending on landholding culture and social structure. The United States has 9 million cows in 75,000 dairy herds, with an average herd size of 120 cows. The number of small herds is falling rapidly with the 3,100 herds with over 500 cows producing 51% of U.S. milk in 2007. The United Kingdom dairy herd overall has nearly 1.5 million cows, with about 100 head reported on an average farm. In New Zealand, the average herd has more than 375 cows, while in Australia, there are approximately 220 cows in the average herd.

Cows on a dairy farm in Maryland, U.S.

Pasteurization is the process of heating milk to a high enough temperature for a short period of time to kill the microbes in the milk and increase keep time and decrease spoilage time by killing the microbes, decrease the transmission of infection and eliminates enzymes that reduce the quality and shelf life Pasteurization is either completed at 63 °C for 30 minutes or a flash pasteurization is completed for 15 seconds at 72 °C.

To maintain lactation, a dairy cow must be bred and produce calves. Depending on market conditions, the cow may be bred with a "dairy bull" or a "beef bull." Female calves (heifers) with dairy breeding may be kept as replacement cows for the dairy herd. If a replacement cow turns out to be a substandard producer of milk, she then goes to market and can be slaughtered for beef. Male calves can either be used later as a breeding bull or sold and used for veal or beef. Dairy farmers usually begin breeding or artificially inseminating heifers around 13 months of age. A cow's gestation period is approximately nine months. Newborn calves are removed from their mothers quickly, usually within three days, as the mother/calf bond intensifies over time and delayed separation can cause extreme stress on both cow and calf.

Domestic cows can live to 20 years; however, those raised for dairy rarely live that long, as the average cow is removed from the dairy herd around age four and marketed for beef. In 2014, approximately 9.5% of the cattle slaughtered in the U.S. were culled dairy cows: cows that can no longer be seen as an economic asset to the dairy farm. These animals may be sold due to reproductive problems or common diseases of milk cows such as mastitis and lameness.

Calf

Market calves are generally sold at two weeks of age and bull calves may fetch a premium over heifers due to their size, either current or potential. Calves may be sold for veal, or for one of several types of beef production, depending on available local crops and markets. Such bull calves may be castrated if turnout onto pastures is envisaged, in order to render the animals less aggressive. Purebred bulls from elite cows may be put into progeny testing schemes to find out whether they might become superior sires for breeding. Such animals may become extremely valuable.

Most dairy farms separate calves from their mothers within a day of birth to reduce transmission of disease and simplify management of milking cows. Studies have been done allowing calves to remain with their mothers for 1, 4, 7 or 14 days after birth. Cows whose calves were removed longer than one day after birth showed increased searching, sniffing and vocalizations. However, calves allowed to remain with their mothers for longer periods showed weight gains at three times the rate of early removals as well as more searching behavior and better social relationships with other calves.

After separation, some young dairy calves subsist on commercial milk replacer, a feed based on dried milk powder. Milk replacer is an economical alternative to feeding whole milk because it is cheaper, can be bought at varying fat and protein percentages, and is typically less contaminated than whole milk when handled properly. Some farms pasteurize and feed calves milk from the cows in the herd instead of using replacer. A day-old calf consumes around 5 liters of milk per day.

Bull

A bull calf with high genetic potential may be reared for breeding purposes. It may be kept by a dairy farm as a herd bull, to provide natural breeding for the herd of cows. A bull may service up to 50 or 60 cows during a breeding season. Any more and the sperm count will decline, leading to cows "returning to service" (to be bred again). A herd bull may only stay for one season since over two years old their temperament becomes too unpredictable.

Bull calves intended for breeding commonly are bred on specialized dairy breeding farms, not production farms. These farms are the major source of stocks for artificial insemination.

Milk Production Levels

Dairy Cows, Collins Center, New York, 1999

A cow will produce large amounts of milk over its lifetime. Certain breeds produce more milk than others; however, different breeds produce within a range of around 6,800 to 17,000 kg (15,000 to 37,500 lbs) of milk per lactation. The Holstein Friesian is the main breed of dairy cattle in Australia, and said to have the "world's highest" productivity, at 10000L of milk per year. The average for a single dairy cow in the US in 2007 was 9164.4 kg (20,204 lbs) per year, excluding milk consumed by her calves, whereas the same average value for a single cow in Israel was reported in the Philippine press to be 12,240 kg in 2009. Production levels peak at around 40 to 60 days after calving. The cow is then bred. Production declines steadily afterwards, until, at about 305 days after calving, the cow is 'dried off', and milking ceases. About sixty days later, one year after the birth of her previous calf, a cow will calve again. High production cows are more difficult to breed at a one-year interval. Many farms take the view that 13 or even 14 month cycles are more appropriate for this type of cow.

Dairy cows may continue to be economically productive for many lactations. In most cases, 10 lactations are possible. The chances of problems arising which may lead to a cow being culled are high, however; the average herd life of US Holstein is today fewer than 3 lactations. This requires more herd replacements to be reared or purchased. Over 90% of all cows are slaughtered for 4 main reasons:

- Infertility - failure to conceive and reduced milk production.

  Cows are at their most fertile between 60 and 80 days after calving. Cows remaining "open" (not with calf) after this period become increasingly difficult to breed, which may be due to poor health. Failure to expel the afterbirth from a previous pregnancy, luteal cysts, or metritis, an infection of the uterus, are common causes of infertility.

- Mastitis - a persistent and potentially fatal mammary gland infection, leading to high somatic cell counts and loss of production.

  Mastitis is recognized by a reddening and swelling of the infected quarter of the udder and the presence of whitish clots or pus in the milk. Treatment is possible with long-acting antibiotics but milk from such cows is not marketable until drug residues have left the cow's system, also called withdrawal period.

- Lameness - persistent foot infection or leg problems causing infertility and loss of production.

  High feed levels of highly digestible carbohydrate cause acidic conditions in the cow's rumen. This leads to Laminitis and subsequent lameness, leaving the cow vulnerable to other foot infections and problems which may be exacerbated by standing in faeces or water soaked areas.

- Production - some animals fail to produce economic levels of milk to justify their feed costs.

  Production below 12 to 15 litres of milk per day is not economically viable.

Cow longevity is strongly correlated with production levels. Lower production cows live longer than high production cows, but may be less profitable. Cows no longer wanted for milk production are sent to slaughter. Their meat is of relatively low value and is generally used for processed meat. Another factor affecting milk production is the stress the cow is faced with. Psychologists at the University of Leicester, UK, analyzed the musical preference of milk cows and found out that music actually influences the dairy cow's lactation. Calming music can improve milk yield, probably because it reduces stress and relaxes the cows in much the same way as it relaxes humans.

Cow Comfort and its Effects on Milk Production

Certain behaviors such as eating, rumination, and laying down can be related to the health of the cow and cow comfort. These behaviors can also be related to the productivity of the cows. Likewise, stress, disease, and discomfort will have a negative effect on the productivity of the dairy cows. Therefore, it can be said that it is in the best interest of the farmer to increase eating, rumination, and laying down and decrease stress, disease, and discomfort to achieve the maximum productivity possible. Also, estrous behaviors such as mounting can be a sign of cow comfort, since if a cow is lame, nutritionally deficient, or are housed in an over crowded barn, the performance of estrous behaviors will be altered.

Feeding behaviors are obviously important for the dairy cow, as feeding is how the cow will ingest dry matter, however, the cow must ruminate to fully digest the feed and utilize the nutrients in the feed. Dairy cows with good rumen health will likely be more profitable than cows with poor rumen health, as a healthy rumen will aid in the digestion of nutrients. An increase in the time a cow spends ruminating is associated with the increase in health and an increase in milk production.

Cows have a high motivation to lay down so farmers should be conscious of this, not only because they have a high motivation to lay down, but also because laying down can increase milk yield. When the lactating dairy cow lays down, blood flow is increased to the mammary gland which in return results in a higher milk yield.

To ensure that the dairy cows lay down as much as needed, the stalls must be comfortable. Put very simply, a stall should have a rubber mat, bedding, and be large enough for the cow to lay down and get up comfortably. Signs that the stalls may not be comfortable enough for the cows are the cows are standing, either ruminating or not, instead of laying down, or perching, which is when the cow has its front end in the stall and their back end out of the stall.

There are 2 types of housing systems in dairy production, free style housing and tie stall. Free style housing is where the cow is free to walk around and interact with its environment and other members of the herd. Tie stall housing is when the cow is chained to a stantion stall with the milking units and feed coming to them.

By-products

By-products of milk include butterfat, cream, curds, and whey. Butterfat is the fat in milk. The cream is the yellowish part of the milk. The cream contains 18–40% butterfat. Whey is the watery part of the milk.

The industry can be divided into 2 market territories; fluid milk and industrialized milk such as yogurt, cheeses, and ice cream.

Reproduction

Since the 1950s, artificial insemination (AI) is used at most dairy farms; these farms may keep no bull. Artificial insemination uses estrus synchronization to indicate when the cow is going through ovulation and is susceptible to fertilization. Advantages of using AI include its low cost and ease compared to maintaining a bull, ability to select from a large number of bulls, elimination of diseases in the dairy industry, improved genetics and improved animal welfare Rather than a large bull jumping on a smaller heifer or weaker cow, AI allows the farmer to complete the breeding procedure within 5 minutes with minimum stress placed on the individual females body

More recently, embryo transfer has been used to enable the multiplication of progeny from elite cows. Such cows are given hormone treatments to produce multiple embryos. These are then 'flushed' from the cow's uterus. 7-12 embryos are consequently removed from these donor cows and transferred into other cows who serve as surrogate mothers. The result will be between 3 and 6 calves instead of the normal single, or rarely, twins.

Hormone Use

Hormone treatments are sometimes given to dairy cows in some countries to increase reproduction and to increase milk production.

The hormones are used to produce multiple embryos have to be administered at specific times to dairy cattle to induce ovulation. Frequently, for economic considerations, these drugs are also used to synchronise a group of cows to ovulate simultaneously. The hormones prostaglandin, gonadotropin-releasing hormone, and progesterone are used for this purpose and sold under the brand names Lutalyse, Cystorelin, Estrumate, Estroplan, Factrel, Prostamate, Fertagyl, Insynch, and Ovacyst. They may be administered by injection.

About 17% of dairy cows in the United States are injected with Bovine somatotropin, also called recombinant bovine somatotropin (rBST), recombinant bovine growth hormone (rBGH), or artificial growth hormone. The use of this hormone increases milk production from 11%–25%. The U.S. Food and Drug Administration (FDA) has ruled that rBST is harmless to people. The use of rBST is banned in Canada, parts of the European Union, as well as Australia and New Zealand.

In Canada, Canadian Dairy farmers have high screening procedures they have to go through every time the milk is retrieved from the farm; if the regulations are not met the milk does not get loaded onto the truck for further processing. There is to be no medication or hormones in the milk for safety reasons

Nutrition

Dairy cattle at feeding time

Nutrition plays an important role in keeping cattle healthy and strong. Implementing an adequate nutrition program can also improve milk production and reproductive performance. Nutrient requirements may not be the same depending on the animal's age and stage of production.

Forages, which refer especially to hay or straw, are the most common type of feed used. Cereal grains, as the main contributors of starch to diets, are important in meeting the energy needs of dairy cattle. Barley is one example of grain that is extensively used around the world. Barley is grown in temperate to subarctic climates, and it is transported to those areas lacking the necessary amounts of grain. Although variations may occur, in general, barley is an excellent source of balanced amounts of protein, energy, and fiber.

Ensuring adequate body fat reserves is essential for cattle to produce milk and also to keep reproductive efficiency. However, if cattle get excessively fat or too thin, they run the risk of developing metabolic problems and may have problems with calving. Scientists have found that a variety of fat supplements can benefit conception rates of lactating dairy cows. Some of these different fats include oleic acids, found in canola oil, animal tallow, and yellow grease; palmitic acid found in granular fats and dry fats; and linolenic acids which are found in cottonseed, safflower, sunflower, and soybean. It is also important to note that proper levels of fat also improve cattle longevity.

Using by-products is one way of reducing the normally high feed costs. However, lack of knowledge of their nutritional and economic value limits their use. Although the reduction of costs may be significant, they have to be used carefully because animal may have negative reactions to radical changes in feeds, (e.g. fog fever). Such a change must then be made slowly and with the proper follow up.

Pesticide Use

A survey of the primary dairy producing areas in the US indicated that 13 percent of lactating animals were treated with insecticides permethrin, pyrethrin, coumaphos, and dichlorvos primarily by daily or every-other-day coat sprays. Workers, particularly in stanchion barns, may be exposed to higher than recommended amounts of these pesticides.

Breeds

According to the Purebred Dairy Cattle Association, PDCA, there are 7 major dairy breeds in the United States. These are: Holstein, Brown Swiss, Guernsey, Ayrshire, Jersey, Red and White, and Milking Shorthorn.

Holstein cows have distinct white and black markings. Holstein cows are the biggest of all U.S. dairy breeds.

A full mature Holstein cow usually weighs around 1,500 pounds and is 58 inches tall at the shoulder. They are known for their outstanding milk production among the main breeds of dairy cattle. An average Holstein cow produces around 23,000 pounds of milk each lactation. Of the 9 million dairy cows in the U.S., approximately 90% of them are of the Holstein descent. The top breed of dairy cow within Canada's national herd category is Holstein, taking up 93% of the dairy cow population, have a production rate of 10 257 kg of milk per cow that contains 3.9% butter fat and 3.2% protein

Brown Swiss cows are widely accepted as the oldest dairy cattle breed, originally coming from a part of northeastern Switzerland. Some experts think that the modern Brown Swiss skeleton is similar to one found that looks to be from around the year 4000 B.C. Also, there is evidence that monks started breeding these cows about 1000 years ago.

The Ayrshire breed first originated in the County of Ayr in Scotland. It became regarded as a well established breed in 1812. The different breeds that were crossed to form the Ayrshire are not exactly known. However, there is evidence that several breeds were crossed with the native cattle to create the breed.

Guernsey cows originated just off the coast of France on the small Isle of Guernsey. The breed was first known as a separate breed around 1700. Guernseys are known for their ability to produce very high quality milk from grass. Also, the term "Golden Guernsey" is very common as Guernsey cattle produce rich, yellow milk rather than the standard white milk other cow breeds produce.

The Jersey breed of dairy cow originated on a small island located off the coast of France called Jersey. Being one of the oldest breeds of dairy cattle they now only occupy 4%

of the Canadian National Herd Being purebred and having genetic improved over 5 centuries Jersey cows, according to available data, have been in the UK area since about the year 1741. When they were in this area, they were not known as Jerseys, but rather as Alderneys. The period between 1860 and around 1914 was the best time for Jerseys. In this time span, many countries other than the United States started importing this breed, including Canada, South Africa, and New Zealand, among others. a Jersey recently won the Senior Champion Female title for Jersey breed at the World Dairy Exposition in Wisconsin, US. Musqie Iatola Martha-ET was produced by Dillmans of Muquodobit in Nova Scotia Among the smallest of the dairy breeds, the average Jersey cow matures at approximately 6699 kg, with a typical weight range between 800 and 1,200 pounds. Jerseys produce milk with approximately 5% butter fat and 3.80% protein. This high fat content means the milk is often used for making ice cream and cheeses. According to the American Jersey Cattle Association, Jerseys are found on 20 percent of all US dairy farms and are the primary breed on about 4 percent of dairies. According to North Dakota State University, the fat content of the Jersey cow is nearly 5 percent—4.9 percent, to be exact. It's also the highest in protein, at 3.8 percent.

Amongst the Bos indicus, the most popular dairy breed in the world is Sahiwal of the Indian subcontinent. It does not give as much milk as the Taurine breeds, but it is by far the most suitable breed for warmer climates. Australian Friesian Sahiwal and Australian Milking Zebu have been developed in Australia using Sahiwal genetics. Gir, another of the Bos Indicus breeds, has been improved in Brazil for its milk production and is widely used there for dairy.

Animal Welfare

Animal welfare refers to both the physical and mental state of an animal, and how it is coping with its situation. An animal is considered in a good state of welfare if it is able to express its innate behaviour, comfortable, healthy, safe, well nourished, and is not suffering from negative states such as distress, fear and pain. Good animal welfare requires disease prevention and veterinary treatment, appropriate shelter, management, nutrition, humane handling, transport and eventually, humane slaughter.

Proper animal handling, or stockmanship, is crucial to dairy animals' welfare as well as the safety of their handlers. Improper handling techniques can stress cattle leading to impaired production and health, such as increased slipping injuries. Additionally, the majority of nonfatal worker injuries on a dairy farm are from interactions with cattle. Dairy animals are handled on a daily basis for a wide variety of purposes including health-related management practices and movement from freestalls to the milking parlor. Due to the prevalence of human-animal interactions on dairy farms, researchers, veterinarians, and farmers alike have focused on furthering our understanding of stockmanship and educating agriculture workers. Stockmanship is a complex concept that involves the timing, positioning, speed, direction of movement, and sounds and touch of the handler. A recent survey of Minnesota dairy farms revealed that 42.6% of workers learned stockmanship techniques from a family members, and 29.9% had participated in stockmanship train-ing. However, as the growing U.S. dairy industry increasingly relies on an immigrant workforce, stockmanship training and education resources will become more pertinent. Clearly communicating and managing a large culturally diverse workforce brings new challenges such as language barriers and time limitations. Organizations like the Upper Midwest Agriculture Safety and Health Center (UMASH) offer resources such as bilin-gual training videos, fact sheets, and informational posters for dairy worker training. Ad-ditionally the Beef Quality Assurance Program offer seminars, live demonstrations, and online resources for stockmanship training.

The practice of dairy production in a factory farm environment has been criticized by animal rights activists. Some of the ethical reasons regarding dairy production cited include how often the dairy cattle are impregnated, the separation of calves from their mothers, and the fact that the cows are considered "spent" and culled at a relatively young age, as well as environmental concerns regarding dairy production.

The production of milk requires that the cow be in lactation, which is a result of the cow having given birth to a calf. The cycle of insemination, pregnancy, parturition, and lactation is followed by a "dry" period of about two months before calving, which allows udder tissue to regenerate. A dry period that falls outside this time frames can result in decreased milk production in subsequent lactation. Dairy operations therefore include both the production of milk and the production of calves. Bull calves are either castrat-ed and raised as steers for beef production or veal.

Animal rights groups such as Mercy for Animals also raise welfare concerns by citing undercover footage showing abusive practices at factory farms.

Silage

Cattle eating corn silage

Silage is fermented, high-moisture stored fodder which can be fed to cattle, sheep and other such ruminants (cud-chewing animals) or used as a biofuel feedstock for anaerobic digesters. It is fermented and stored in a process called *ensilage, ensiling* or *silaging*, and is usually made from grass crops, including maize, sorghum or other cereals, using the entire green plant (not just the grain). Silage can be made from many field crops, and special terms may be used depending on type (*oatlage* for oats, *haylage* for alfalfa – but see below for the different British use of the term *haylage*).

Silage is made either by placing cut green vegetation in a silo or pit, by piling it in a large heap and compressing it down so as to leave as little oxygen as possible and then covering it with a plastic sheet, or by wrapping large round bales tightly in plastic film.

Making Silage

Partially dried mown grass is formed into cylindrical bales in the field...

...and are then sealed in polywrap.

The crops suitable for ensilage are the ordinary grasses, clovers, alfalfa, vetches, oats, rye and maize; various weeds may also be stored in silos, notably spurrey such as *Spergula arvensis*. Silage must be made from plant material with a suitable moisture content, about 50% to 60% depending on the means of storage, the degree of compression, and the amount of water that will be lost in storage, but not exceeding 75%. Weather during harvest need not be as fair and dry as when harvesting for drying. For corn, harvest begins when the whole-plant moisture is at a suitable level, ideally a few days before it is ripe. For pasture-type crops, the grass is mowed and allowed to wilt for a day or so until the moisture content drops to a suitable level. Ideally the crop is mowed when in full flower, and deposited in the silo on the day of its cutting.

After harvesting, crops are shredded to pieces about 0.5 in (1.3 cm) long. The material is spread in uniform layers over the floor of the silo, and closely packed. When the silo is filled or the stack built, a layer of straw or some other dry porous substance may be spread over the surface. In the silo the pressure of the material, when chaffed, excludes air from all but the top layer; in the case of the stack extra pressure is applied by weights in order to prevent excessive heating.

Equipment

Forage harvesters collect and chop the plant material, and deposit it in trucks or wagons. These forage harvesters can be either tractor-drawn or self-propelled. Harvesters blow the chaff into the wagon through a chute at the rear or side of the machine. Chaff may also be emptied into a bagger, which puts the silage into a large plastic bag that is laid out on the ground.

MB Trac rolling a silage heap or "clamp" in Victoria, Australia

In North America, Australia, northwestern Europe, and frequently in New Zealand, silage is placed in large heaps on the ground and rolled by tractor to push out the air, then wrapped in plastic covers held down by reused tires or tire ring walls.

In New Zealand and Northern Europe, the silo or "pit" is often a bunker built into the side of a bank, usually made out of concrete or old wooden railroad ties (railway sleepers). The chopped grass can then be dumped in at the top, to be drawn from the bottom in winter. This requires considerable effort to compress the stack in the silo to cure it properly. Again, the pit is covered with plastic sheet and weighed down with tire weights.

In an alternative method, the cut vegetation is baled, making *balage* (North America) or *silage bales* (UK). The grass or other forage is cut and partly dried until it contains 30–40% moisture (much drier than bulk silage, but too damp to be stored as dry hay). It is then made into large bales which are wrapped tightly in plastic to exclude air. The plastic may wrap the whole of each cylindrical or cuboid bale, or be wrapped around only the curved sides of a cylindrical bale, leaving the ends uncovered. In this case, the bales are placed tightly end to end on the ground, making a long continuous "sausage" of silage, often at the side of a field. The wrapping may be performed by a bale wrapper, while the baled silage is handled using a bale handler or a front-loader, either impaling the bale on a flap, or by using a special grab. The flaps do not hole the bales.

In the UK, baled silage is most often made in round bales about 4 feet by 4 feet, individually wrapped with four to six layers of "bale wrap plastic" (black, white or green 25 micrometre stretch film). The dry matter can vary a lot but can be from about 20% dry matter upwards. The continuous "sausage" referred to above is made with a special machine which wraps the bales as they are pushed through a rotating hoop which applies the bale wrap to the outside of the bales (round or square) in a continuous wrap. The machine places the bales on the ground after wrapping by moving forward slowly during the wrapping process (search for "tube liner" various makes).

Haylage is a name for high dry matter silage of around 45% to 75%. Horse haylage is usually 55% to 75% dry matter, made in small bales or larger bales. Handling of wrapped bales is most often with some type of gripper that squeezes the plastic-covered bale between two metal parts to avoid puncturing the plastic. Simple fixed versions are available for round bales which are made of two shaped pipes or tubes spaced apart to slide under the sides of the bale, but when lifted will not let it slip through. Often used on the tractor rear three-point linkage, they incorporate a trip tipping mechanism which can flip the bales over on to the flat side/end for storage on the thickest plastic layers.

Fermentation

Silage undergoes anaerobic fermentation, which starts about 48 hours after the silo is filled, and converts sugars to acids. Fermentation is essentially complete after about two weeks.

Before anaerobic fermentation starts, there is an aerobic phase in which the trapped oxygen is consumed. The closeness with which the fodder is packed, determines the nature of the resulting silage by regulating the chemical reactions that occur in the stack. When closely packed, the supply of oxygen is limited; and the attendant acid fermentation brings about decomposition of the carbohydrates present into acetic, butyric and lactic acids. This product is named sour silage. If, on the other hand, the fodder is unchaffed and loosely packed, or the silo is built gradually, oxidation proceeds more rapidly and the temperature rises; if the mass is compressed when the temperature is 140 to 160 Fahrenheit or 60 to 71 ° Centigrade, the action ceases and sweet silage results. The nitrogenous ingredients of the fodder also suffer change: in making sour silage as much as one-third of the albuminoids may be converted into amino and ammonium compounds; while in making sweet silage a smaller proportion is changed, but they become less digestible. If the fermentation process is poorly managed, sour silage acquires an unpleasant odour due to excess production of ammonia or butyric acid (the latter is responsible for the smell of rancid butter).

In the past, the fermentation was conducted by indigenous microorganisms, but, today, some bulk silage is inoculated with specific microorganisms to speed fermentation or improve the resulting silage. Silage inoculants contain one or more strains of lactic acid bacteria, and the most common is *Lactobacillus plantarum*. Other bacteria used in inoculants include *Lactobacillus buchneri*, *Enterococcus faecium* and *Pediococcus* species.

Optimum Grass to Grow for Quality Silage

Ryegrasses have high sugars and respond to nitrogen fertiliser better than any other grass species. These two qualities have made ryegrass the most popular grass for silage making for the last sixty years. There are three ryegrasses in seed form and commonly used: Italian, Perennial and Hybrid.

Pollution and Waste

The fermentation process of silo or pit silage releases liquid. Silo effluent is corrosive. It can also contaminate water sources unless collected and treated. The high nutrient content can lead to eutrophication (hypertrophication), growth of bacterial or algal blooms.

Plastic sheeting used for sealing pit or baled silage needs proper disposal, and some areas have recycling schemes for it. Traditionally, farms have burned silage plastics; however odor and smoke concerns have led certain communities to restrict that practice.

Storing Silage

Silage underneath plastic sheeting is held down by scrap tires. Concrete beneath the silage prevents liquor leaching out.

Silage must be firmly packed to minimize the oxygen content, or it will spoil. Silage goes through four major stages in a silo:

- Presealing, which, after the first few days after filling a silo, enables some respiration and some dry matter (DM) loss, but stops

- Fermentation, which occurs over a few weeks; pH drops; there is more DM loss, but hemicellulose is broken down; aerobic respiration stops

- Infiltration, which enables some oxygen infiltration, allowing for limited microbial respiration; available carbohydrates (CHOs) are lost as heat and gas

- Emptying, which exposes surface, causing additional loss; rate of loss increases.

Anaerobic Digestion

Anaerobic digesters

Silage is a useful feedstock for anaerobic digestion. Here silage can be fed into anaerobic digesters to produce biogas that, in turn, can be used to generate electricity and heat.

Safety

Silos are potentially hazardous: deaths may occur in the process of filling and maintaining them, and several safety precautions are necessary. There is a risk of injury by machinery or from falls. When a silo is filled, fine dust particles in the air can become explosive because of their large aggregate surface area. Also, fermentation presents respiratory hazards. The ensiling process produces "silo gas" during the early stages of the fermentation process. Silage gas contains nitric oxide (NO), which will react with oxygen (O_2) in the air to form nitrogen dioxide (NO_2), which is toxic. Lack of oxygen inside the silo can cause asphyxiation. Molds that grow when air reaches cured silage can cause organic dust toxic syndrome. Collapsing silage from large bunker silos has caused deaths. Silage itself poses no special danger.

Nutrition

Ensilage can be substituted for root crops. Bulk silage is commonly fed to dairy cattle, while baled silage tends to be used for beef cattle, sheep and horses. The advantages of silage as animal feed are several:

- During fermentation, the silage bacteria act on the cellulose and carbohydrates in the forage to produce volatile fatty acids (VFAs), such as acetic, propionic, lactic, and butyric acids. By lowering pH, these create a hostile environment for competing bacteria that might cause spoilage. The VFAs thus act as natural preservatives, in the same way that the lactic acid in yogurt and cheese increases the preservability of what began as milk, or vinegar (dilute acetic acid) preserves pickled vegetables. This preservative action is particularly important during winter in temperate regions, when green forage is unavailable.

- When silage is prepared under optimal conditions, the modest acidity also has the effect of improving palatability and provides a dietary contrast for the animal. (However, excessive production of acetic and butyric acids can reduce palatability: the mix of bacteria is ideally chosen so as to maximize lactic acid production.)

- Several of the fermenting organisms produce vitamins: for example, lactobacillus species produce folic acid and vitamin B12.

- The fermentation process that produces VFA also yields energy that the bacteria use: some of the energy is released as heat. Silage is thus modestly lower in caloric content than the original forage, in the same way that yoghurt has modestly fewer calories than milk. However, this loss of energy is offset by the preservation characteristics and improved digestibility of silage.

History

Using the same technique as the process for making sauerkraut, green fodder was preserved for animals in parts of Germany since the start of the 19th century. This gained the attention of a French agriculturist, Auguste Goffart of Sologne, near Orléans, who published a book in 1877 which described the experiences of preserving green crops in silos. Goffart's experience attracted considerable attention. The conditions of dairy farming in the USA suited the ensiling of green corn fodder, and was soon adopted by New England farmers. Francis Morris of Maryland prepared the first silage produced in America in 1876. The favourable results obtained in the U.S. led to the introduction of the system in the United Kingdom, where Thomas Kirby first introduced the process for British dairy herds.

Early silos were made of stone or concrete either above or below ground, but it is recognized that air may be sufficiently excluded in a tightly pressed stack, though in this case a few inches of the fodder round the sides is generally useless owing to mildew. In the U.S. structures were typically constructed of wooden cylinders to 35 or 40 ft. in depth.

In the early days of mechanized agriculture, stalks were cut and collected manually using a knife and horsedrawn wagon, and fed into a stationary machine called a "silo filler" that chopped the stalks and blew them up a narrow tube to the top of a tower silo.

Dairy

Old mountain pasture dairy in Schröcken, Vorarlberg, Austria, in the Bregenz Forest

A dairy is a business enterprise established for the harvesting or processing (or both) of animal milk – mostly from cows or goats, but also from buffaloes, sheep, horses, or camels – for human consumption. A dairy is typically located on a dedicated dairy farm or in a section of a multi-purpose farm (mixed farm) that is concerned with the harvesting of milk.

Terminology differs between countries. For example, in the United States, the entire dairy farm is commonly called a "dairy." The building or farm area where milk is harvested from the cow is often called a "milking parlor" or "parlor." The farm area where milk is stored in bulk tanks is known as the farm's "milk house." Milk is then hauled (usually by truck) to a "dairy plant," also referred to as a "dairy", where raw milk is further processed and prepared for commercial sale of dairy products. In New Zealand, farm areas for milk harvesting are also called "milking parlours", and are historically known as "milking sheds." As in the United States, sometimes milking sheds are referred to by their type, such as "herring bone shed" or "pit parlour". Parlour design has evolved from simple barns or sheds to large rotary structures in which the workflow (throughput of cows) is very efficiently handled. In some countries, especially those with small numbers of animals being milked, the farm may perform the functions of a dairy plant, processing their own milk into salable dairy products, such as butter, cheese, or yogurt. This on-site processing is a traditional method of producing specialist milk products, common in Europe.

In the United States a *dairy* can also be a place that processes, distributes and sells dairy products, or a room, building or establishment where milk is stored and processed into milk products, such as butter or cheese. In New Zealand English the singular use of the word *dairy* almost exclusively refers to a corner shop, or superette. This usage is historical as such shops were a common place for the public to buy milk products.

As an attributive, the word *dairy* refers to milk-based products, derivatives and processes, and the animals and workers involved in their production: for example dairy cattle, dairy goat. A dairy farm produces milk and a dairy factory processes it into a variety of dairy products. These establishments constitute the global dairy industry, a component of the food industry.

History

Milk producing animals have been domesticated for thousands of years. Initially, they were part of the subsistence farming that nomads engaged in. As the community moved about the country, their animals accompanied them. Protecting and feeding the animals were a big part of the symbiotic relationship between the animals and the herders.

In the more recent past, people in agricultural societies owned dairy animals that they milked for domestic and local (village) consumption, a typical example of a cottage industry. The animals might serve multiple purposes (for example, as a draught animal for pulling a plough as a youngster, and at the end of its useful life as meat). In this case the animals were normally milked by hand and the herd size was quite small, so that all of the animals could be milked in less than an hour—about 10 per milker. These tasks were performed by a *dairymaid* (*dairywoman*) or *dairyman*. The word *dairy* harkens back to Middle English *dayerie*, *deyerie*, from *deye* (female servant or dairymaid) and further back to Old English *dæge* (kneader of bread).

With industrialisation and urbanisation, the supply of milk became a commercial industry, with specialised breeds of cattle being developed for dairy, as distinct from beef or draught animals. Initially, more people were employed as milkers, but it soon turned to mechanisation with machines designed to do the milking.

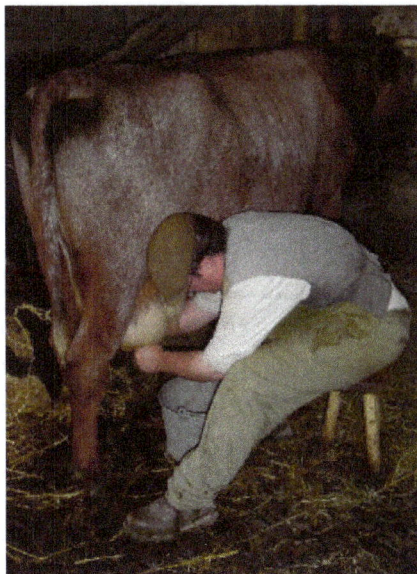
Farmer milking a cow by hand

Historically, the milking and the processing took place close together in space and time: on a dairy farm. People milked the animals by hand; on farms where only small numbers are kept, hand-milking may still be practiced. Hand-milking is accomplished by grasping the teats (often pronounced *tit* or *tits*) in the hand and expressing milk either by squeezing the fingers progressively, from the udder end to the tip, or by squeezing the teat between thumb and index finger, then moving the hand downward from udder towards the end of the teat. The action of the hand or fingers is designed to close off the milk duct at the udder (upper) end and, by the movement of the fingers, close the duct progressively to the tip to express the trapped milk. Each half or quarter of the udder is emptied one milk-duct capacity at a time.

The *stripping* action is repeated, using both hands for speed. Both methods result in the milk that was trapped in the milk duct being squirted out the end into a bucket that is supported between the knees (or rests on the ground) of the milker, who usually sits on a low stool.

Traditionally the cow, or cows, would stand in the field or paddock while being milked. Young stock, heifers, would have to be trained to remain still to be milked. In many countries, the cows were tethered to a post and milked. The problem with this method is that it relies on quiet, tractable beasts, because the hind end of the cow is not restrained.

In 1937, it was found that bovine somatotropin (BST or bovine growth hormone) would increase the yield of milk. Monsanto Company developed a synthetic (recombinant) version of this hormone (rBST). In February 1994, rBST was approved by the Food and Drug Administration (FDA) for use in the U.S. It was common in the U.S., but has lost popularity due to consumer demands for rBST-free cows. Only about 25% of dairy cows receive rBST anymore.

However, there are claims that this practice can have negative consequences for the animals themselves. A European Union scientific commission was asked to report on the incidence of mastitis and other disorders in dairy cows, and on other aspects of the welfare of dairy cows. The commission's statement, subsequently adopted by the European Union, stated that the use of rBST substantially increased health problems with cows, including foot problems, mastitis and injection site reactions, impinged on the welfare of the animals and caused reproductive disorders. The report concluded that on the basis of the health and welfare of the animals, rBST should not be used. Health Canada prohibited the sale of rBST in 1999; the recommendations of external committees were that, despite not finding a significant health risk to humans, the drug presented a threat to animal health and, for this reason, could not be sold in Canada.

Structure of the Industry

While most countries produce their own milk products, the structure of the dairy industry varies in different parts of the world. In major milk-producing countries most milk is distributed through whole sale markets. In Ireland and Australia, for example, farmers' co-operatives own many of the large-scale processors, while in the United States many farmers and processors do business through individual contracts. In the United States, the country's 196 farmers' cooperatives sold 86% of milk in the U.S. in 2002, with five cooperatives accounting for half that. This was down from 2,300 cooperatives in the 1940s. In developing countries, the past practice of farmers marketing milk in their own neighborhoods is changing rapidly. Notable developments include considerable foreign investment in the dairy industry and a growing role for dairy co-operatives. Output of milk is growing rapidly in such countries and presents a major source of income growth for many farmers.

Wawa Dairy Farms in Pennsylvania

As in many other branches of the food industry, dairy processing in the major dairy producing countries has become increasingly concentrated, with fewer but larger and more efficient plants operated by fewer workers. This is notably the case in the United States, Europe, Australia and New Zealand. In 2009, charges of anti-trust violations have been made against major dairy industry players in the United States, which critics call Big Milk. Another round of price fixing charges was settled in 2016.

Government intervention in milk markets was common in the 20th century. A limited anti-trust exemption was created for U.S. dairy cooperatives by the Capper-Volstead Act of 1922. In the 1930s, some U.S. states adopted price controls, and Federal Milk Marketing Orders started under the Agricultural Marketing Agreement Act of 1937 and continue in the 2000s. The Federal Milk Price Support Program began in 1949. The Northeast Dairy Compact regulated wholesale milk prices in New England from 1997 to 2001.

Plants producing liquid milk and products with short shelf life, such as yogurts, creams and soft cheeses, tend to be located on the outskirts of urban centres close to consumer markets. Plants manufacturing items with longer shelf life, such as butter, milk powders, cheese and whey powders, tend to be situated in rural areas closer to the milk supply. Most large processing plants tend to specialise in a limited range of products.

Exceptionally, however, large plants producing a wide range of products are still common in Eastern Europe, a holdover from the former centralized, supply-driven concept of the market under Communist governments.

As processing plants grow fewer and larger, they tend to acquire bigger, more automated and more efficient equipment. While this technological tendency keeps manufacturing costs lower, the need for long-distance transportation often increases the environmental impact.

Milk production is irregular, depending on cow biology. Producers must adjust the mix of milk which is sold in liquid form vs. processed foods (such as butter and cheese) depending on changing supply and demand.

Farming

A rotary dairy shed

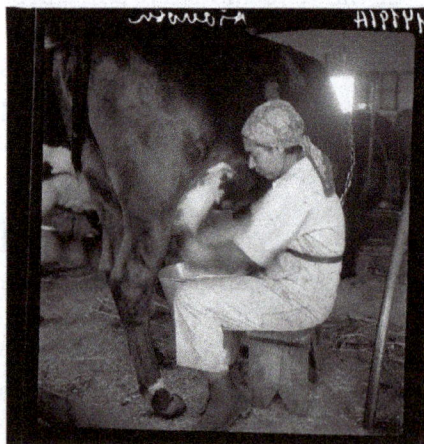
A cow being milked in Israel

When it became necessary to milk larger cows, the cows would be brought to a shed or barn that was set up with bails (milking stalls) where the cows could be confined while they were milked. One person could milk more cows this way, as many as 20 for a skilled worker. But having cows standing about in the yard and shed waiting to be milked is not good for the cow, as she needs as much time in the paddock grazing as is possible. It is usual to restrict the twice-daily milking to a maximum of an hour and a half each time. It makes no difference whether one milks 10 or 1000 cows, the milking time should not exceed a total of about three hours each day for any cow.

As herd sizes increased there was more need to have efficient milking machines, sheds, milk-storage facilities (vats), bulk-milk transport and shed cleaning capabilities and the means of getting cows from paddock to shed and back.

As herd numbers increased so did the problems of animal health. In New Zealand two approaches to this problem have been used. The first was improved veterinary medicines (and the government regulation of the medicines) that the farmer could use. The other was the creation of *veterinary clubs* where groups of farmers would employ a veterinarian (vet) full-time and share those services throughout the year. It was in the vet's interest to keep the animals healthy and reduce the number of calls from farmers, rather than to ensure that the farmer needed to call for service and pay regularly.

This daily milking routine goes on for about 300 to 320 days per year that the cow stays in milk. Some small herds are milked once a day for about the last 20 days of the production cycle but this is not usual for large herds. If a cow is left unmilked just once she is likely to reduce milk-production almost immediately and the rest of the season may see her *dried off* (giving no milk) and still consuming feed. However, once-a-day milking is now being practised more widely in New Zealand for profit and lifestyle reasons. This is effective because the fall in milk yield is at least partially offset by labour and cost savings from milking once per day. This compares to some intensive farm systems in the United States that milk three or more times per day due to higher milk yields per cow and lower marginal labor costs.

Farmers who are contracted to supply liquid milk for human consumption often have to manage their herd so that the contracted number of cows are in milk the year round, or the required minimum milk output is maintained. This is done by mating cows outside their natural mating time so that the period when each cow in the herd is giving maximum production is in rotation throughout the year.

Northern hemisphere farmers who keep cows in barns almost all the year usually manage their herds to give continuous production of milk so that they get paid all year round. In the southern hemisphere the cooperative dairying systems allow for two months on no productivity because their systems are designed to take advantage of maximum grass and milk production in the spring and because the milk processing plants pay bonuses in the dry (winter) season to carry the farmers through the mid-winter break from milking. It also means that cows have a rest from milk production when they are most heavily pregnant. Some year-round milk farms are penalised financially for over-production at any time in the year by being unable to sell their overproduction at current prices.

Artificial insemination (AI) is common in all high-production herds.

Industrial Processing

A Fonterra cooperative dairy factory in Australia

Interior of a cheese factory in Seine-et-Marne, France

Dairy plants process the raw milk they receive from farmers so as to extend its marketable life. Two main types of processes are employed: heat treatment to ensure the safety of milk for human consumption and to lengthen its shelf-life, and dehydrating dairy products such as butter, hard cheese and milk powders so that they can be stored.

Cream and Butter

Today, milk is separated by huge machines in bulk into cream and skim milk. The cream is processed to produce various consumer products, depending on its thickness, its suitability for culinary uses and consumer demand, which differs from place to place and country to country.

Some cream is dried and powdered, some is condensed (by evaporation) mixed with varying amounts of sugar and canned. Most cream from New Zealand and Australian factories is made into butter. This is done by churning the cream until the fat globules coagulate and form a monolithic mass. This butter mass is washed and, sometimes, salted to improve keeping qualities. The residual buttermilk goes on to further processing. The butter is packaged (25 to 50 kg boxes) and chilled for storage and sale. At a later stage these packages are broken down into home-consumption sized packs.

Skimmed Milk

The product left after the cream is removed is called skim, or skimmed, milk. To make a consumable liquid a portion of cream is returned to the skim milk to make *low fat milk* (semi-skimmed) for human consumption. By varying the amount of cream returned, producers can make a variety of low-fat milks to suit their local market. Other products, such as calcium, vitamin D, and flavouring, are also added to appeal to consumers.

Casein

Casein is the predominant phosphoprotein found in fresh milk. It has a very wide range of uses from being a filler for human foods, such as in ice cream, to the manufacture of products such as fabric, adhesives, and plastics.

Cheese

Cheese is another product made from milk. Whole milk is reacted to form curds that can be compressed, processed and stored to form cheese. In countries where milk is legally allowed to be processed without pasteurization, a wide range of cheeses can be made using the bacteria naturally in the milk. In most other countries, the range of cheeses is smaller and the use of artificial cheese curing is greater. Whey is also the byproduct of this process. Some people with lactose intolerance are surprisingly able to eat certain types of cheese. This is because some traditionally made hard cheeses, and soft ripened cheeses may create less reaction than the equivalent amount of milk because of the processes involved. Fermentation and higher fat content contribute to lesser amounts of lactose. Traditionally made Emmental or Cheddar might contain 10% of the lactose found in whole milk. In addition, the aging methods of traditional cheeses (sometimes over two years) reduce their lactose content to practically nothing. Commercial cheeses, however, are often manufactured by processes that do not have the same lactose-reducing properties. Ageing of some cheeses is governed by regulations; in other cases there is no quantitative indication of degree of ageing and concomitant lactose reduction, and lactose content is not usually indicated on labels.

Whey

In earlier times, whey or milk serum was considered to be a waste product and it was, mostly, fed to pigs as a convenient means of disposal. Beginning about 1950, and mostly since about 1980, lactose and many other products, mainly food additives, are made from both casein and cheese whey.

Yogurt

Yogurt (or yoghurt) making is a process similar to cheese making, only the process is arrested before the curd becomes very hard.

Milk Powders

Milk is also processed by various drying processes into powders. Whole milk, skim milk, buttermilk, and whey products are dried into a powder form and used for human and animal consumption. The main difference between production of powders for human or for animal consumption is in the protection of the process and the product from contamination. Some people drink milk reconstituted from powdered milk, because milk is about 88% water and it is much cheaper to transport the dried product.

Other Milk Products

Kumis is produced commercially in Central Asia. Although it is traditionally made from mare's milk, modern industrial variants may use cow's milk instead.

Milking

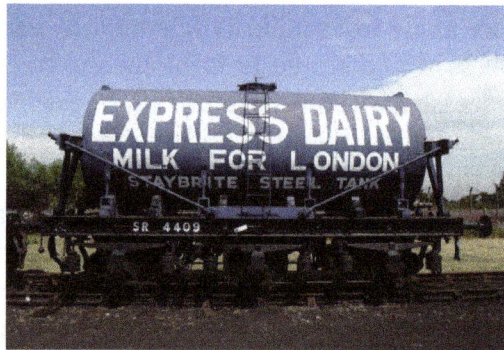

Preserved Express Dairies three-axle Milk Tank Wagon at the Didcot Railway Centre, based on an SR chassis

Milk churns on a railway platform

Originally, milking and processing took place on the dairy farm itself. Later, cream was separated from the milk by machine on the farm, and transported to a factory to be made into butter. The skim milk was fed to pigs. This allowed for the high cost of transport (taking the smallest volume high-value product), primitive trucks and the poor quality of roads. Only farms close to factories could afford to take whole milk, which was essential for cheesemaking in industrial quantities, to them.

Originally milk was distributed in 'pails', a lidded bucket with a handle. These proved impractical for transport by road or rail, and so the milk churn was introduced, based on the tall conical shape of the butter churn. Later large railway containers, such as the British Railway Milk Tank Wagon were introduced, enabling the transport of larger quantities of milk, and over longer distances.

The development of refrigeration and better road transport, in the late 1950s, has meant that most farmers milk their cows and only temporarily store the milk in large refrigerated bulk tanks, from where it is later transported by truck to central processing facilities.

In many European countries, particularly the United Kingdom, milk is then delivered direct to customers' homes by a milk float.

Milking Machines

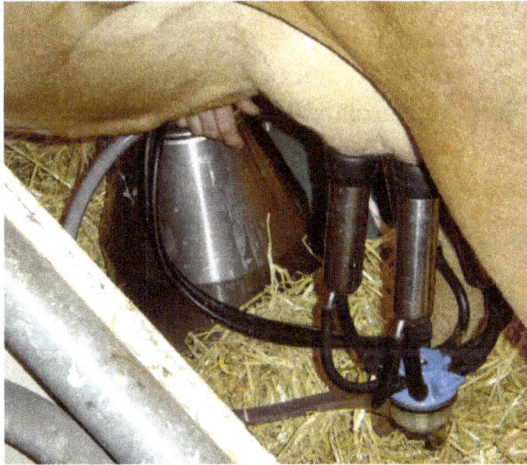
The milking machine extracts milk from all teats.

Milking machines are used to harvest milk from cows when manual milking becomes inefficient or labour-intensive. One early model was patented in 1907. The milking unit is the portion of a milking machine for removing milk from an udder. It is made up of a claw, four teatcups, (Shells and rubber liners) long milk tube, long pulsation tube, and a pulsator. The claw is an assembly that connects the short pulse tubes and short milk tubes from the teatcups to the long pulse tube and long milk tube. (Cluster assembly) Claws are commonly made of stainless steel or plastic or both. Teatcups are composed of a rigid outer shell (stainless steel or plastic) that holds a soft inner liner or *inflation*. Transparent sections in the shell may allow viewing of liner collapse and milk flow. The annular space between the shell and liner is called the pulse chamber.

Milking machines work in a way that is different from hand milking or calf suckling. Continuous vacuum is applied inside the soft liner to massage milk from the teat by creating a pressure difference across the teat canal (or opening at the end of the teat).

Vacuum also helps keep the machine attached to the cow. The vacuum applied to the teat causes congestion of teat tissues (accumulation of blood and other fluids). Atmospheric air is admitted into the pulsation chamber about once per second (the pulsation rate) to allow the liner to collapse around the end of teat and relieve congestion in the teat tissue. The ratio of the time that the liner is open (milking phase) and closed (rest phase) is called the pulsation ratio.

The four streams of milk from the teatcups are usually combined in the claw and transported to the milkline, or the collection bucket (usually sized to the output of one cow) in a single milk hose. Milk is then transported (manually in buckets) or with a combination of airflow and mechanical pump to a central storage vat or bulk tank. Milk is refrigerated on the farm in most countries either by passing through a heat-exchanger or in the bulk tank, or both.

The photo to the right shows a bucket milking system with the stainless steel bucket visible on the far side of the cow. The two rigid stainless steel teatcup shells applied to the front two quarters of the udder are visible. The top of the flexible liner is visible at the top of the shells as are the short milk tubes and short pulsation tubes extending from the bottom of the shells to the claw. The bottom of the claw is transparent to allow observation of milk flow. When milking is completed the vacuum to the milking unit is shut off and the teatcups are removed.

Milking machines keep the milk enclosed and safe from external contamination. The interior 'milk contact' surfaces of the machine are kept clean by a manual or automated washing procedures implemented after milking is completed. Milk contact surfaces must comply with regulations requiring food-grade materials (typically stainless steel and special plastics and rubber compounds) and are easily cleaned.

Most milking machines are powered by electricity but, in case of electrical failure, there can be an alternative means of motive power, often an internal combustion engine, for the vacuum and milk pumps.

Milking Shed Layouts

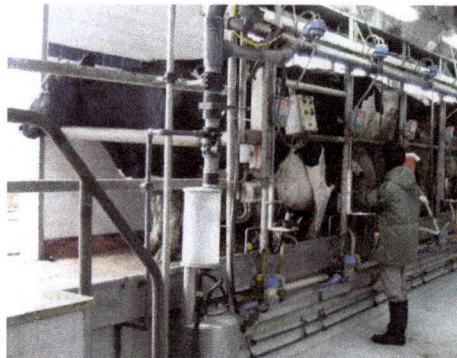

Milking parlour at Pardes Hanna Agricultural High School, Israel

Bail-style Sheds

This type of milking facility was the first development, after open-paddock milking, for many farmers. The building was a long, narrow, *lean-to* shed that was open along one long side. The cows were held in a yard at the open side and when they were about to be milked they were positioned in one of the bails (stalls). Usually the cows were restrained in the bail with a breech chain and a rope to restrain the outer back leg. The cow could not move about excessively and the milker could expect not to be kicked or trampled while sitting on a (three-legged) stool and milking into a bucket. When each cow was finished she backed out into the yard again. The UK bail, initially developed by Wiltshire dairy farmer Arthur Hosier, was a six standing mobile shed with steps that the cow mounted, so the herdsman didn't have to bend so low. The milking equipment was much as today, a vacuum from a pump, pulsators, a claw-piece with pipes leading to the four shells and liners that stimulate and suck the milk from the teat. The milk went into churns, via a cooler.

As herd sizes increased a door was set into the front of each bail so that when the milking was done for any cow the milker could, after undoing the leg-rope and with a remote link, open the door and allow her to exit to the pasture. The door was closed, the next cow walked into the bail and was secured. When milking machines were introduced bails were set in pairs so that a cow was being milked in one paired bail while the other could be prepared for milking. When one was finished the machine's cups are swapped to the other cow. This is the same as for *Swingover Milking Parlours* as described below except that the cups are loaded on the udder from the side. As herd numbers increased it was easier to double-up the cup-sets and milk both cows simultaneously than to increase the number of bails. About 50 cows an hour can be milked in a shed with 8 bails by one person. Using the same teat cups for successive cows has the danger of transmitting infection, mastitis, from one cow to another. Some farmers have devised their own ways to disinfect the clusters between cows.

Herringbone Milking Parlours

In herringbone milking sheds, or parlours, cows enter, in single file, and line up almost perpendicular to the central aisle of the milking parlour on both sides of a central pit in which the milker works (you can visualise a fishbone with the ribs representing the cows and the spine being the milker's working area; the cows face outward). After washing the udder and teats the cups of the milking machine are applied to the cows, from the rear of their hind legs, on both sides of the working area. Large herringbone sheds can milk up to 600 cows efficiently with two people.

Swingover Milking Parlours

Swingover parlours are the same as herringbone parlours except they have only one set of milking cups to be shared between the two rows of cows, as one side is being milked the cows on the other side are moved out and replaced with unmilked ones. The advantage of this system is that it is less costly to equip, however it operates at slightly better than half-speed and one would not normally try to milk more than about 100 cows with one person.

80-stand rotary dairy that is fully computerised and records milk production

Rotary Milking Sheds

Rotary milking sheds (also known as Rotary milking parlor) consist of a turntable with about 12 to 100 individual stalls for cows around the outer edge. A "good" rotary will be operated with 24–32 (~48–50+) stalls by one (two) milkers. The turntable is turned by an electric-motor drive at a rate that one turn is the time for a cow to be milked completely. As an empty stall passes the entrance a cow steps on, facing the center, and rotates with the turntable. The next cow moves into the next vacant stall and so on. The operator, or milker, cleans the teats, attaches the cups and does any other feeding or whatever husbanding operations that are necessary. Cows are milked as the platform rotates. The milker, or an automatic device, removes the milking machine cups and the cow backs out and leaves at an exit just before the entrance. The rotary system is capable of milking very large herds—over a thousand cows.

Automatic Milking Sheds

Automatic milking or 'robotic milking' sheds can be seen in Australia, New Zealand, the U.S., Canada, and many European countries. Current automatic milking sheds use the voluntary milking (VM) method. These allow the cows to voluntarily present themselves for milking at any time of the day or night, although repeat visits may be limited by the farmer through computer software. A robot arm is used to clean teats and apply milking equipment, while automated gates direct cow traffic, eliminating the need for the farmer to be present during the process. The entire process is computer controlled.

Supplementary Accessories in Sheds

Farmers soon realised that a milking shed was a good place to feed cows supplementary foods that overcame local dietary deficiencies or added to the cows' wellbeing and production. Each bail might have a box into which such feed is delivered as the cow arrives so that she is eating while being milked. A computer can read the eartag of each animal to ration the correct individual supplement. A close alternative is to use 'out-of-parlour-feeders', stalls that respond to a transponder around the cow's neck that is programmed to provide each cow with a supplementary feed, the quantity dependent on her production, stage in lactation, and the benefits of the main ration

The holding yard at the entrance of the shed is important as a means of keeping cows moving into the shed. Most yards have a powered gate that ensures that the cows are kept close to the shed.

Water is a vital commodity on a dairy farm: cows drink about 20 gallons (80 litres) a day, sheds need water to cool and clean them. Pumps and reservoirs are common at milking facilities. Water can be warmed by heat transfer with milk.

Temporary Milk Storage

Milk coming from the cow is transported to a nearby storage vessel by the airflow leaking around the cups on the cow or by a special "air inlet" (5-10 l/min free air) in the claw. From there it is pumped by a mechanical pump and cooled by a heat exchanger. The milk is then stored in a large vat, or bulk tank, which is usually refrigerated until collection for processing.

Waste Disposal

Manure spreader going to the field from a dairy farm, Elba, New York.

In countries where cows are grazed outside year-round, there is little waste disposal to deal with. The most concentrated waste is at the milking shed, where the animal waste may be liquefied (during the water-washing process) or left in a more solid form, either to be returned to be used on farm ground as organic fertilizer.

In the associated milk processing factories, most of the waste is washing water that is treated, usually by composting, and spread on farm fields in either liquid or solid form.

This is much different from half a century ago, when the main products were butter, cheese and casein, and the rest of the milk had to be disposed of as waste (sometimes as animal feed).

In dairy-intensive areas, various methods have been proposed for disposing of large quantities of milk. Large application rates of milk onto land, or disposing in a hole, is problematic as the residue from the decomposing milk will block the soil pores and thereby reduce the water infiltration rate through the soil profile. As recovery of this effect can take time, any land based application needs to be well managed and considered. Other waste milk disposal methods commonly employed include solidification and disposal at a solid waste landfill, disposal at a wastewater treatment plant, or discharge into a sanitary sewer.

Associated Diseases

Dairy products manufactured under unsanitary or unsuitable conditions have an increased chance of containing bacteria. Proper sanitation practices help to reduce the rate of bacterial contamination, and pasteurization greatly decreases the amount of contaminated milk that reaches the consumer. Many countries have required government oversight and regulations regarding dairy production, including requirements for pasteurization.

- Leptospirosis is an infection that can be transmitted to people who work in dairy production through exposure to urine or to contaminated water or soil.

- Cowpox is a virus that today is rarely found in either cows or humans. It is a historically important disease, as it led to the first vaccination against the now eradicated smallpox.

- Tuberculosis is able to be transmitted from cattle mainly via milk products that are unpasteurised. The disease has been eradicated from many countries by testing for the disease and culling suspected animals.

- Brucellosis is a bacterial disease transmitted to humans by dairy products and direct animal contact. Brucellosis has been eradicated from certain countries by testing for the disease and culling suspected animals.

- Listeria is a bacterial disease associated with unpasteurised milk, and can affect some cheeses made in traditional ways. Careful observance of the traditional cheesemaking methods achieves reasonable protection for the consumer.

- Crohn's disease has been linked to infection with the bacterium *M. paratuberculosis*, which has been found in pasteurized retail milk in the UK and the USA. *M. paratuberculosis* causes a similar disorder, Johne's disease, in livestock.

Animal Welfare

A portion of the population, including many vegans and Jains, object to dairy production as unethical, cruel to animals, and environmentally deleterious. They do not consume dairy products. They state that cattle suffer under conditions employed by the dairy industry.

Dairy Product

Dairy products are derived from milk

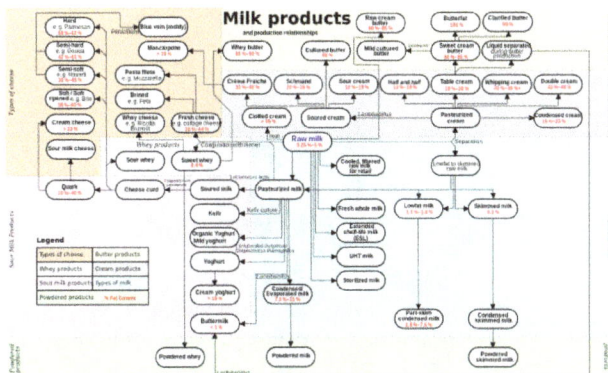

Milk products and production relationships

Dairy products or milk products are a type of food produced from or containing the milk of mammals, primarily cattle, water buffaloes, goats, sheep, and camels. Dairy products include food items like yogurt, cheese, and butter. A facility that produces dairy products is a dairy or dairy factory. Dairy products are often consumed worldwide, except for most of East and Southeast Asia and parts of central Africa.

Types of Dairy Products

A dairy farm

A selection of three common dairy products made by a South African dairy company: a box of full cream, long life milk, a bottle of strawberry drinking yogurt, and a carton of passion fruit yogurt

The milk products of the Water buffaloes (super carabaos, Philippine Carabao Center)

- Milk after optional homogenization, pasteurization, in several grades after stan-dardization of the fat level, and possible addition of the bacteria *Streptococcus lactis* and *Leuconostoc citrovorum*

  - *Crème fraîche*, slightly fermented cream

    - Clotted cream, thick, spoonable cream made by heating milk

- Single cream, double cream and whipping cream

- *Smetana*, Central and Eastern European variety of sour cream

  o Cultured milk resembling buttermilk, but uses different yeast and bacterial cultures

  o Kefir, fermented milk drink from the Northern Caucasus

  o *Kumis/Airag*, slightly fermented mares' milk popular in Central Asia

  o Powdered milk (or milk powder), produced by removing the water from (usually skim) milk

    - Whole milk products

    - Buttermilk products

    - Skim milk

    - Whey products

    - High milk-fat and nutritional products (for infant formulas)

    - Cultured and confectionery products

  o Condensed milk, milk which has been concentrated by evaporation, with sugar added for reduced process time and longer life in an opened can

  o *Khoa*, milk which has been completely concentrated by evaporation, used in Indian cuisine including gulab jamun, peda, etc.)

  o Evaporated milk, (less concentrated than condensed) milk without added sugar

  o Ricotta, acidified whey, reduced in volume

  o Infant formula, dried milk powder with specific additives for feeding human infants

  o Baked milk, a variety of boiled milk that has been particularly popular in Russia

- Butter, mostly milk fat, produced by churning cream

  o Buttermilk, the liquid left over after producing butter from cream, often dried as livestock feed

  o *Ghee*, clarified butter, by gentle heating of butter and removal of the solid matter

  o *Smen*, a fermented, clarified butter used in Moroccan cooking

- o   Anhydrous milkfat (clarified butter)
- Cheese, produced by coagulating milk, separating from whey and letting it ripen, generally with bacteria and sometimes also with certain molds
    - o   Curds, the soft, curdled part of milk (or skim milk) used to make cheese
    - o   *Paneer*
    - o   Whey, the liquid drained from curds and used for further processing or as a livestock feed
    - o   Cottage cheese
    - o   Quark
    - o   Cream cheese, produced by the addition of cream to milk and then curdled to form a rich curd or cheese
    - o   *Fromage frais*
- Casein are
    - o   Caseinates, sodium or calcium salts of casein
    - o   Milk protein concentrates and isolates
    - o   Whey protein concentrates and isolates, reduced lactose whey
    - o   Hydrolysates, milk treated with proteolytic enzymes to alter functionality
    - o   Mineral concentrates, byproduct of demineralizing whey
- Yogurt, milk fermented by *Streptococcus salivarius* ssp. *thermophilus* and *Lactobacillus delbrueckii* ssp. *bulgaricus* sometimes with additional bacteria, such as *Lactobacillus acidophilus*
    - o   *Doogh*
    - o   *Lassi*, Indian subcontinent
    - o   *Leben*
- Clabber, milk naturally fermented to a yogurt-like state
- Gelato, slowly frozen milk and water, lesser fat than ice cream
- Ice cream, slowly frozen cream, milk, flavors and emulsifying additives (dairy ice cream)
    - o   Ice milk, low-fat version of ice cream

o   Frozen custard

o   Frozen yogurt, yogurt with emulsifiers

- Other

o   *Viili*

o   *Kajmak*

o   *Filmjölk*

o   *Piimä*

o   *Vla*

o   *Dulce de leche*

o   *Skyr*

o   Junket, milk solidified with rennet

Health

Dairy products can cause health issues for individuals who have lactose intolerance or a milk allergy.

Additionally dairy products including cheese, ice cream, milk, butter, and yogurt can contribute significant amounts of cholesterol and saturated fat to the diet. Diets high in fat and especially in saturated fat can increase the risk of heart disease and can cause other serious health problems. However, it has been shown that there is no connection between dairy consumption (excluding butter) and cardiovascular disease, even though dairy tends to be higher in saturated fats.

There is no excess cardiovascular risk with dietary calcium intake but calcium supplements are associated with a higher risk of coronary artery calcification. Anderson JJ, Kruszka B, Delaney JA, et al. Calcium intake from diet and supplements and the risk of coronary artery calcification and its progression among older adults: 10-year follow-up of the Multi-Ethnic Study of Atherosclerosis (MESA). J Am Heart Assoc 20161; DOI:10.1161/jaha.116.003815

Consumption Patterns Worldwide

Rates of dairy consumption vary widely worldwide. High-consumption countries consume over 150 kg per capita per year: Argentina, Armenia, Australia, Costa Rica, Europe, Israel, Kyrgyzstan, North America and Pakistan. Medium-consumption countries consume 30 to 150 kg per capita per year: India, Iran, Japan, Kenya, Mexico, Mongolia, New Zealand, North and Southern Africa, most of the Middle East, and most of Latin America and the Caribbean. Low-consumption countries consume under 30 kg per capita per year: Senegal, most of Central Africa, and most of East and Southeast Asia.

Avoidance

Some groups avoid dairy products for non-health related reasons:

- Religious - Some religions restrict or do not allow for the consumption of dairy products. For example, some scholars of Jainism advocate not consuming any dairy products because dairy is perceived to involve violence against cows. Strict Judaism requires that meat and dairy products not be served at the same meal, served or cooked in the same utensils, or stored together, as prescribed in Deuteronomy 14:21.

- Ethical - Veganism is the avoidance of all animal products, including dairy products, most often due to the ethics regarding how dairy products are produced. The ethical reasons for avoiding dairy include how dairy is produced, how the animals are handled, and the environmental effect of dairy production.

References

- Sisney, Jason; Garosi, Justin. "California is the Leading Farm State". Legislative Analyst's Office. Retrieved 28 May 2016.

- "dairy product - definition of dairy product in English | Oxford Dictionaries". Oxford Dictionaries | English. Retrieved 2016-11-11.

- writer, Julie R. Thomson Food (2014-10-14). "PSA: Eggs Are NOT Dairy". The Huffington Post. Retrieved 2016-11-11.

- "Is Butter a Dairy Product, and Does it Contain Lactose?". Authority Nutrition. 2016-07-01. Retrieved 2016-11-11.

- U.S. Department of Agriculture, National Agriculture Statistics Service (April 2015). "Livestock Slaughter Annual Summary" (PDF). Archived from the original (PDF) on 12 September 2015. Retrieved 2015-11-24.

- MacDonald, James; Newton, Doris (1 December 2014). "Milk Production Continues Shifting to Large-Scale Farms". Amber Waves. United States Department of Agriculture. Retrieved 24 March 2015.

- "Purebred Dairy Cattle Association". Purebred Dairy Cattle Association. Purebred Dairy Cattle Association. Retrieved 17 September 2014.

- McLean, Amy. "Donkey milk for human health?". Tri-State Livestock News. Swift Communications, Inc. Retrieved 28 December 2013.

Cream and Butter: Primary Dairy Products

Cream is a product that is produced by the dairy. Cream is also dried and then made into a powder and sold into the market. The primary dairy products that are used on a daily basis are sour cream, crème fraiche, butter and churning. The major categories of cream and butter are dealt with great details in the following chapter.

Cream

A bottle of unhomogenised milk, with the cream clearly visible, resting on top of the milk.

Cream is a dairy product composed of the higher-butterfat layer skimmed from the top of milk before homogenization. In un-homogenized milk, the fat, which is less dense, will eventually rise to the top. In the industrial production of cream, this process is accelerated by using centrifuges called "separators". In many countries, cream is sold in several grades depending on the total butterfat content. Cream can be dried to a powder for shipment to distant markets. Cream has high levels of saturated fat.

Cream skimmed from milk may be called "sweet cream" to distinguish it from whey cream skimmed from whey, a by-product of cheese-making. Whey cream has a lower fat content and tastes more salty, tangy and "cheesy". In many countries, cream is usually sold partially fermented: sour cream, crème fraîche, and so on.

Cream has many culinary uses in sweet, bitter, salty and tangy dishes.

Cream produced by cattle (particularly Jersey cattle) grazing on natural pasture often contains some natural carotenoid pigments derived from the plants they eat; this gives the cream a slight yellow tone, hence the name of the yellowish-white color, cream. This is also the origin of butter's yellow color. Cream from goat's milk, or from cows fed indoors on grain or grain-based pellets, is white.

Cuisine

A slice of pumpkin pie topped with a whipped cream rose

Cream is used as an ingredient in many foods, including ice cream, many sauces, soups, stews, puddings, and some custard bases, and is also used for cakes. Whipped cream is served as a topping on ice cream sundaes, milkshakes, egg nog and sweet pies. Irish cream is an alcoholic liqueur which blends cream with whiskey, and often honey, wine, or coffee. Cream is also used in Indian curries such as masala dishes.

Cream (usually light/single cream or half and half) is often added to coffee in the US and Canada.

Both single and double cream can be used in cooking. Double cream or full-fat crème fraîche are often used when cream is added to a hot sauce, to prevent any problem with it separating or "splitting". Double cream can be thinned with milk to make an approximation of single cream.

The French word *crème* denotes not only dairy cream, but also other thick liquids such as sweet and savory custards, which are normally made with milk, not cream.

Types

Stewed nectarines and heavy cream

Different grades of cream are distinguished by their fat content, whether they have been heat-treated, whipped, and so on. In many jurisdictions, there are regulations for each type.

United States

In the United States, cream is usually sold as:

Name	Fat Content	Main Uses
Half and Half	10.5-18%	To whiten coffee (and tea).
Light Cream	18-30%	Also called "table" cream or "coffee" cream. Old style product for whitening coffee and also as an enriching ingredient in sauces and other recipes. This product is becoming difficult to find at retail in many areas.
Whipping Cream	30-36%	Generally 33%. Used in sauces and soups and as a pourable or whipped garnish. Whipping will only attain soft peaks. Some products labeled "Whipping Cream" contain small amounts of gelatin as an added stabilizer for improved whipping.
Heavy (Whipping) Cream	36% min.	For whipping when pert stable peaks are desired. Also used as a luxurious pourable garnish on fresh fruit and hot cereals.
Manufacturer's Cream	>=40%	Used in commercial and professional production applications. Not generally available at retail until recently.

Most cream products sold in the United States at retail contain the minimum permissible fat content for their product type, e.g., "Half and half" almost always contains only 10.5% butterfat. Not all grades are defined by all jurisdictions, and the exact fat content ranges vary. The above figures are based on the Code of Federal Regulations, Title 21, Part 131

Australia

The Australia New Zealand Food Standards Code - Standard 2.5.2 - Defines cream as milk product comparatively rich in fat, in the form of an emulsion of fat-in-skim milk, which can be obtained by separation from milk. Cream must contain no less than 350 g/kg of milk fat.

Manufacturers labels may distinguish between different fat contents, a general guideline is as follows:

Name	Fat Content	Main Uses
Extra light (or 'lite')	12-12.5%	
Light (or 'lite')	18-20%	
Thickened Cream	35-36.5%	with added gelatine and/or other thickeners to give the cream a creamier texture, also possibly with stabilizers to aid the consistency of whipped cream (this would be the cream to use for whipped cream, not necessarily for cooking)
Single Cream	~ 35%	Recipes calling for 'single cream' are referring to pure or thickened cream with about 35% fat.
Double Cream	48-60%	

UK

In the United Kingdom, the types of cream are legally defined as followed:

Name	Minimum milk fat	Additional definition	Main uses
Clotted cream	55%	is heat-treated	Served as it is. A traditional part of a cream tea.
Extra-thick double cream	48%	is heat-treated then quickly cooled	Thickest available fresh cream, spooned onto pies, puddings, and desserts (cannot be poured due to its consistency)
Double cream	48%		Whips easily and thickest for puddings and desserts, can be piped once whipped
Whipping cream	35%		Whips well but lighter, can be piped once whipped
Whipped cream	35%	has been whipped	Decorations on cakes, topping for ice cream, fruit and so on.

Sterilized cream	23%	is sterilized	
Cream or single cream	18%	is not sterilized	Poured over puddings, used in sauces
Sterilized half cream	12%	is sterilized	
Half cream			Uncommon, some cocktails

Meiji whipping cream

Canada

Canadian cream definitions are similar to those used in the United States, except for that of "light cream". In Canada, "light cream" is very low-fat cream, usually with 5% or 6% butterfat. Specific product characteristics are generally uniform throughout Canada, but names vary by both geographic and linguistic area and by manufacturer. It can be quite confusing: "coffee cream" may be 10% or 18% and "half-and-half" ("crème légère") may be 3%, 5%, 6% or 10%, all depending on location and brand.

Name	Minimum milk fat	Additional definition	Main uses
Manufacturing cream	40%	Crème fraîche is also 40% - 45% but is an acidified cultured product rather than sweet cream.	Commercial production.
Whipping cream	33%—36%	Also as cooking or "thick" cream 35% with added stabilizers. Heavy cream must be at least 36%. In Francophone areas: crème à fouetter 35%; and for cooking, crème à cuisson 35%, crème à l'ancienne 35% or crème épaisse 35%.	Whips into a creamy and smooth topping that is used for pastries, fresh fruits, desserts, hot cocoa, etc. Cooking version is formulated to resist breaking when heated (as in sauces).

Table cream	15%—18%	Coffee cream. Also as cooking or "thick" cream 15% with added stabilizers. In Francophone areas: crème de table 15% or crème à café 18%; and for cooking, crème champêtre 15%, crème campagnarde (country cream) 15% or crème épaisse 15%.	Added as rich whitener to coffee. Ideal for soups, sauces and veloutés. Garnishing fruit and desserts. Cooking version is formulated to resist breaking when heated.
Half and half	10%	Cereal cream (at least one producer calls it coffee cream; another calls it Creamo™ light cream). Product with the most butterfat in the light cream category. In Francophone areas: crème à café 10% and sometimes crème légère 10%.	Poured over hot cereal as a garnish. Ideal in sauces for vegetables, fish, meat, poultry, and pasta. Also in cream soups.
Light cream	3%—10%	LIght cream 6%. In Francophone areas: mélange de lait et de crème pour café 5%, Crémette™ 5% or crème légère 3% to 10%. An admixture of milk and cream.	5% product is similar to the richest Guernsey or Jersey milk. A lower fat alternative to table cream in coffee.

Japan

In Japan, cream sold in supermarkets is usually between 35% and 48% butterfat.

Russia

Russia, as well as other EAC countries, legally separates cream into two classes: normal (10-34% butterfat) and heavy (35-58%), but the industry has pretty much standardized around the following types:

English	Russian	Transliteration	Milk fat (wt%)
Low-fat or drinking cream	Нежирные (питьевые) сливки	Nezhirnÿe (pityevÿe) slivki	10%
(Normal) Cream	Сливки	Slivki	15% or 20%
Whipping cream	Сливки для взбивания	Slivki dlya vzbivaniya	33% or 35%
Double cream	Двойные (жирные) сливки	Dvoinÿe (Zhirnÿe) slivki	48%

Switzerland

In Switzerland, the types of cream are legally defined as follows:

English	German	French	Italian	Typical milk fat wt%	Minimum milk fat wt%
Double cream	Doppelrahm	double-crème	doppia panna	45%	45%
Full cream Whipping cream Cream	Vollrahm Schlagrahm Rahm Sahne	crème entière crème à fouetter crème	panna intera panna da montare panna	35%	35%
Half cream	Halbrahm	demi-crème	mezza panna	25%	15%
Coffee cream	Kaffeerahm	crème à café	panna da caffè	15%	15%

Sour cream and crème fraîche (German: Sauerrahm, Crème fraîche; French: crème acidulée, crème fraîche; Italian: panna acidula, crème fraîche) are defined as cream soured by bacterial cultures.

Thick cream (German: verdickter Rahm; French: crème épaissie; Italian: panna addensata) is defined as cream thickened using thickening agents.

Sweden

In Sweden, cream is usually sold as:

- Matlagningsgrädde ("*cooking cream*"), 10-15 %

- Kaffegrädde ("*Coffee cream*"), 10%

- Vispgrädde (*whipping cream*), 36 - 40%

Mellangrädde (27%) is, nowadays, a less common variant. Gräddfil and Creme Fraiche are two common sour cream products.

Processing and Additives

Cream may have thickening agents and stabilizers added. Thickeners include sodium alginate, carrageenan, gelatine, sodium bicarbonate, tetrasodium pyrophosphate, and alginic acid.

Other processing may be carried out. For example, cream has a tendency to produce oily globules (called "feathering") when added to coffee. The stability of the cream may be increased by increasing the non-fat solids content, which can be done by partial demineralisation and addition of sodium caseinate, although this is expensive.

Other Cream Products

Butter is made by churning cream to separate the butterfat and buttermilk. This can be done by hand or by machine.

Whipped cream is made by whisking or mixing air into cream with more than 30% fat, to turn the liquid cream into a soft solid. Nitrous oxide may also be used to make whipped cream.

Sour cream, common in many countries including the U.S., Canada and Australia, is cream (12 to 16% or more milk fat) that has been subjected to a bacterial culture that produces lactic acid (0.5%+), which sours and thickens it.

Crème fraîche (28% milk fat) is slightly soured with bacterial culture, but not as sour or as thick as sour cream. Mexican crema (or cream espesa) is similar to crème fraîche.

Smetana is a heavy cream product (15-40% milk fat) Central and Eastern European sweet or sour cream.

Rjome or rømme is Norwegian sour cream containing 35% milk fat, similar to Icelandic sýrður rjómi.

Clotted cream, common in the United Kingdom, is made through a process that starts by slowly heating whole milk to produce a very high-fat (55%) product. This is similar to Indian malai.

Other Items Called "Cream"

Some non-edible substances are called creams due to their consistency: shoe cream is runny, unlike regular waxy shoe polish; hand/body 'creme' or "skin cream" is meant for moisturizing the skin.

Regulations in many jurisdictions restrict the use of the word *cream* for foods. Words such as *creme, kreme, creame,* or *whipped topping* (e.g., Cool Whip) are often used for products which cannot legally be called cream. Oreo cookies are a type of sandwich cookie in which two biscuits have a soft, sweet filling between them which is called "crème filling". In some cases foods can be described as cream although they do not contain predominantly milk fats; for example in Britain "ice cream" does not have to be a dairy product (although it must be labelled "contains non-milk fat"), and salad cream is the customary name for a condiment that has been produced since the 1920s.

Sour Cream

Sour cream is a dairy product obtained by fermenting regular cream with certain kinds of lactic acid bacteria. The bacterial culture, which is introduced either deliberately or naturally, sours and thickens the cream. Its name comes from the production of lactic acid by bacterial fermentation, which is called souring.

Bowl of chili with sour cream and cheese

Traditional

Traditionally, sour cream was made by letting cream that was skimmed off the top of milk ferment at a moderate temperature. The bacteria that developed during fermentation thickened the cream and made it more acidic, a natural way of preserving it.

Traditional sour cream contains from 18 to 20 percent butterfat and gets its characteristic tang from the lactic acid created by the bacteria.

Commercial Varieties

Commercially produced sour cream usually contains 14 percent milk fat, and can contain additional thickening agents such as gelatin, rennet, guar gum and carrageenan, as well as acids to artificially sour the product.

Light, or reduced-fat, sour cream contains less butterfat than regular sour cream, because it is made from a mixture of milk and cream rather than just cream. Fat-free "sour cream" contains no cream at all, and is made primarily from non-fat milk, modified cornstarch, thickeners and flavoring agents.

Sour cream is not fully fermented, and like many dairy products, must be refrigerated unopened and after use. It is sold with an expiration date stamped on the container, though whether this is a "sell by", a "best by" or a "use by" date varies with local regulation. Refrigerated unopened sour cream can last for 1–2 weeks beyond its *sell by date* while refrigerated opened sour cream generally lasts for 7–10 days.

Uses

Sour cream is used primarily in the cuisines of Europe and North America, often as a condiment. It is a traditional topping for baked potatoes, added cold along with chopped fresh chives. It is used as the base for some creamy salad dressings and can also be used in baking, added to the mix for cakes, cookies, American-style biscuits, doughnuts and scones. It can be eaten as a dessert, with fruits or berries and sugar topping. Also, it is sometimes used on top of waffles in addition to strawberry jam. In Central America, *crema* (a variation of sour cream) is a staple ingredient of a full breakfast.

Sour cream can also provide the base for various forms of dip used for dipping potato chips or crackers, such as onion dip.

In Tex-Mex cuisine, it is often used as a substitute for crema in nachos, tacos, burritos, taquitos or guacamole.

Sour cream is one of the main ingredients in chicken paprikash and beef Stroganoff. It is also the main ingredient in the traditional Norwegian porridge Rømmegrøt which is named after the cream.

A variation of sour cream known as smetana is also popular in eastern Europe and Russia as soup condiment, usually added to individual bowls and stirred until dissolved.

Crème Fraîche

Crème fraîche is a soured cream containing 10–45% butterfat and having a pH of around 4.5. It is soured with bacterial culture, but is less sour than U.S.-style sour cream, and has a lower viscosity and a higher fat content. European labeling regulation disallows any ingredients other than cream and bacterial culture.

Strawberries and crème fraîche

Chilled asparagus soup with crème fraîche and pink peppercorn

The name "crème fraîche" is French, but similar soured creams are found in much of northern Europe.

Terminology

In French-speaking countries, *crème fraîche* may refer to either the thick fermented product, *crème fraîche épaisse* or *fermentée*, or to liquid cream, *crème fraîche liquide* or *fleurette*. In these countries, *crème fraîche* without qualification normally refers to liquid cream, with the thick form usually called *crème épaisse*. In other countries, however, *crème fraîche* without qualification usually refers to the thick, fermented product.

Production

Crème fraîche is produced by adding a starter culture to heavy cream, and allowing it to stand at appropriate temperature until thick. The culture is made up of a mix of bacteria including *Lactococcus* species *L. cremoris*, *L. lactis*, and *L. lactis* biovar *diacetylactis*. This is what gives it the taste that distinguishes it from similar dairy products like sour cream.

In some places in Europe, the fat content of crème fraîche is regulated, and it may not contain ingredients other than cream and starter culture. In North America and the UK, products labeled "low-fat crème fraîche", with about 15% butterfat and with added stabilizers such as xanthan gum or maize/corn starch are commercialized. It is less stable when heated.

History

The crème fraîche from Normandy is famous, and the crème fraîche from a defined area around the town of Isigny-sur-Mer in the Calvados department of Normandy is highly regarded. It is the only cream to have an appellation d'origine contrôlée (AOC), which was awarded in 1986. It is also produced in many other parts of France, with large quantities coming from the major dairy regions of Brittany, Poitou-Charente, Lorraine and Champagne-Ardenne.

Raspberries with crème fraîche and sugar

Uses

Crème fraîche is used both hot and cold in French cuisine. It is often used to finish hot savory sauces; with its fat content greater than 30%, curdling is not a problem. It is also the basis of many desserts and dessert sauces.

Similar Products

Crema Mexicana is a somewhat similar cultured sour cream. *Smetana* from Eastern Europe and Russia is very similar.

Butter

Butter displayed in a market

Butter is a solid dairy product made by churning fresh or fermented cream or milk, to separate the butterfat from the buttermilk. It is generally used as a spread on plain or toasted bread products and a condiment on cooked vegetables, as well as in cooking, such as baking, sauce making, and pan frying. Butter consists of butterfat, milk proteins and water.

Most frequently made from cows' milk, butter can also be manufactured from the milk of other mammals, including sheep, goats, buffalo, and yaks. Salt such as dairy salt, flavorings and preservatives are sometimes added to butter. Rendering butter produces clarified butter or *ghee*, which is almost entirely butterfat.

Butter is a water-in-oil emulsion resulting from an inversion of the cream; in a water-in-oil emulsion, the milk proteins are the emulsifiers. Butter remains a solid when refrigerated, but softens to a spreadable consistency at room temperature, and melts to a thin liquid consistency at 32–35 °C (90–95 °F). The density of butter is 911 g/L (0.950 lbs per US pint).

It generally has a pale yellow color, but varies from deep yellow to nearly white. Its unmodified color is dependent on the animals' feed and is commonly manipulated with food colorings in the commercial manufacturing process, most commonly annatto or carotene.

Etymology

Butter is often served for spreading on bread with a butter knife.

Nevertheless, the earliest attested form of the second stem, *turos* ("cheese"), is the Mycenaean Greek *tu-ro*, written in Linear B syllabic script. The root word persists in the name butyric acid, a compound found in rancid butter and dairy products such as Parmesan cheese.

In general use, the term "butter" refers to the spread dairy product when unqualified by other descriptors. The word commonly is used to describe puréed vegetable or seed and nut products such as peanut butter and almond butter. It is often applied to spread fruit products such as apple butter. Fats such as cocoa butter and shea butter that remain solid at room temperature are also known as "butters". In addition to the act of applying butter being called "to butter", non-dairy items that have a dairy butter consistency may use "butter" to call that consistency to mind, including food items such as maple butter and witch's butter and nonfood items such as baby bottom butter, hyena butter, and rock butter.

Production

Churning cream into butter using a hand-held mixer.

Unhomogenized milk and cream contain butterfat in microscopic globules. These globules are surrounded by membranes made of phospholipids (fatty acid emulsifiers) and proteins, which prevent the fat in milk from pooling together into a single mass. Butter is produced by agitating cream, which damages these membranes and allows the milk fats to conjoin, separating from the other parts of the cream. Variations in the production method will create butters with different consistencies, mostly due to the butterfat composition in the finished product. Butter contains fat in three separate forms: free butterfat, butterfat crystals, and undamaged fat globules. In the finished product, different proportions of these forms result in different consistencies within the butter; butters with many crystals are harder than butters dominated by free fats.

Churning produces small butter grains floating in the water-based portion of the cream. This watery liquid is called buttermilk—although the buttermilk most common today is instead a directly fermented skimmed milk. The buttermilk is drained off; sometimes more buttermilk is removed by rinsing the grains with water. Then the grains are "worked": pressed and kneaded together. When prepared manually, this is done using wooden boards called scotch hands. This consolidates the butter into a solid mass and breaks up embedded pockets of buttermilk or water into tiny droplets.

Commercial butter is about 80% butterfat and 15% water; traditionally made butter may have as little as 65% fat and 30% water. Butterfat is a mixture of triglyceride, a triester derived from glycerol and three of any of several fatty acid groups. Butter becomes rancid when these chains break down into smaller components, like butyric acid and diacetyl. The density of butter is 0.911 g/cm³ (0.527 oz/in³), about the same as ice.

In some countries, butter is given a grade before commercial distribution.

Types

Before modern factory butter making, cream was usually collected from several milkings and was therefore several days old and somewhat fermented by the time it was made into butter. Butter made from a fermented cream is known as cultured butter. During fermentation, the cream naturally sours as bacteria convert milk sugars into lactic acid. The fermentation process produces additional aroma compounds, including diacetyl, which makes for a fuller-flavored and more "buttery" tasting product.[p35] Today, cultured butter is usually made from pasteurized cream whose fermentation is produced by the introduction of *Lactococcus* and *Leuconostoc* bacteria.

Chart of milk products and production relationships, including butter.

Another method for producing cultured butter, developed in the early 1970s, is to produce butter from fresh cream and then incorporate bacterial cultures and lactic acid. Using this method, the cultured butter flavor grows as the butter is aged in cold storage. For manufacturers, this method is more efficient, since aging the cream used to make butter takes significantly more space than simply storing the finished butter product. A method to make an artificial simulation of cultured butter is to add lactic acid and flavor compounds directly to the fresh-cream butter; while this more efficient process is claimed to simulate the taste of cultured butter, the product produced is not cultured but is instead flavored.

Dairy products are often pasteurized during production to kill pathogenic bacteria and other microbes. Butter made from pasteurized fresh cream is called sweet cream butter. Production of sweet cream butter first became common in the 19th century, with the development of refrigeration and the mechanical cream separator. Butter made from fresh or cultured unpasteurized cream is called raw cream butter. While butter made from pasteurized cream may keep for several months, raw cream butter has a shelf life of roughly ten days.

Throughout continental Europe, cultured butter is preferred, while sweet cream butter dominates in the United States and the United Kingdom. Cultured butter is sometimes labeled "European-style" butter in the United States, although cultured butter is made and sold by some, especially Amish, dairies. Commercial raw cream butter is virtually unheard-of in the United States. Raw cream butter is generally only found made at home by consumers who have purchased raw whole milk directly from dairy farmers, skimmed the cream themselves, and made butter with it. It is rare in Europe as well.[p34]

Several "spreadable" butters have been developed. These remain softer at colder temperatures and are therefore easier to use directly out of refrigeration. Some methods modify the makeup of the butter's fat through chemical manipulation of the finished product, some manipulate the cattle's feed, and some incorporate vegetable oil into the butter. "Whipped" butter, another product designed to be more spreadable, is aerated by incorporating nitrogen gas—normal air is not used to avoid oxidation and rancidity.

All categories of butter are sold in both salted and unsalted forms. Either granular salt or a strong brine are added to salted butter during processing. In addition to enhanced flavor, the addition of salt acts as a preservative. The amount of butterfat in the finished product is a vital aspect of production. In the United States, products sold as "butter" must contain at least 80% butterfat. In practice, most American butters contain slightly more than that, averaging around 81% butterfat. European butters generally have a higher ratio—up to 85%.

Liquid clarified butter

Clarified butter is butter with almost all of its water and milk solids removed, leaving almost-pure butterfat. Clarified butter is made by heating butter to its melting point and then allowing it to cool; after settling, the remaining components separate by density. At the top, whey proteins form a skin, which is removed. The resulting butterfat is then poured off from the mixture of water and casein proteins that settle to the bottom.

Ghee is clarified butter that has been heated to around 120 °C (250 °F) after the water evaporated, turning the milk solids brown. This process flavors the ghee, and also produces antioxidants that help protect it from rancidity. Because of this, ghee can keep for six to eight months under normal conditions.

Whey Butter

Butter made in a barn; German painting by Jan Spanjaert.

Cream may be separated (usually by a centrifugal separator) from whey instead of milk, as a byproduct of cheese-making. Whey butter may be made from whey cream. Whey cream and butter have a lower fat content and taste more salty, tangy and "cheesy". They are also cheaper than "sweet" cream and butter. The fat content of whey is low, so 1000 pounds of whey will typically give 3 pounds of butter.

European Butters

There are several butters produced in Europe with Protected geographical indications; these include:

- Beurre d'Ardenne, from Belgium

- Beurre d'Isigny, from France

- Beurre Charentes-Poitou (Which also includes: Beurre des Charentes and Beurre des Deux-Sèvres under the same classification), from France

- Beurre Rose, from Luxembourg

- Mantequilla de Soria, from Spain

- Mantega de l'Alt Urgell i la Cerdanya, from Spain

History

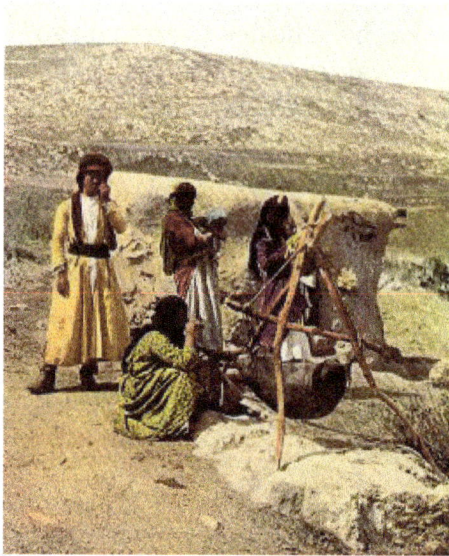

Traditional butter-making in Palestine. Ancient techniques were still practiced in the early 20th century. *National Geographic,* March 1914.

The earliest butter would have been from sheep or goat's milk; cattle are not thought to have been domesticated for another thousand years. An ancient method of butter making, still used today in parts of Africa and the Near East, involves a goat skin half filled with milk, and inflated with air before being sealed. The skin is then hung with ropes on a tripod of sticks, and rocked until the movement leads to the formation of butter.

In the Mediterranean climate, unclarified butter spoils quickly— unlike cheese, it is not a practical method of preserving the nutrients of milk. The ancient Greeks and Romans seemed to have considered butter a food fit more for the northern barbarians. A play by the Greek comic poet Anaxandrides refers to Thracians as *boutyrophagoi,* "butter-eaters". In his *Natural History,* Pliny the Elder calls butter "the most delicate of food among barbarous nations", and goes on to describe its medicinal properties. Later, the physician Galen also described butter as a medicinal agent only.

Historian and linguist Andrew Dalby says most references to butter in ancient Near Eastern texts should more correctly be translated as ghee. Ghee is mentioned in the Periplus of the Erythraean Sea as a typical trade article around the first century CE Arabian Sea, and Roman geographer Strabo describes it as a commodity of Arabia and Sudan. In India, ghee has been a symbol of purity and an offering to the gods—especially Agni, the Hindu god of fire—for more than 3000 years; references to ghee's sacred nature appear numerous times in the *Rigveda,* circa 1500–1200 BCE. The tale of the child Krishna stealing butter remains a popular children's story in India today. Since India's prehistory, ghee has been both a staple food and used for ceremonial purposes, such as fueling holy lamps and funeral pyres.

Middle Ages

Woman churning butter; *Compost et Kalendrier des Bergères*, Paris, 1499

In the cooler climates of northern Europe, people could store butter longer before it spoiled. Scandinavia has the oldest tradition in Europe of butter export trade, dating at least to the 12th century. After the fall of Rome and through much of the Middle Ages, butter was a common food across most of Europe—but had a low reputation, and so was consumed principally by peasants. Butter slowly became more accepted by the upper class, notably when the early 16th century Roman Catholic Church allowed its consumption during Lent. Bread and butter became common fare among the middle class, and the English, in particular, gained a reputation for their liberal use of melted butter as a sauce with meat and vegetables.

In antiquity, butter was used for fuel in lamps as a substitute for oil. The *Butter Tower* of Rouen Cathedral was erected in the early 16th century when Archbishop Georges d'Amboise authorized the burning of butter instead of oil, which was scarce at the time, during Lent.

Across northern Europe, butter was sometimes treated in a manner unheard-of today: it was packed into barrels (firkins) and buried in peat bogs, perhaps for years. Such "bog butter" would develop a strong flavor as it aged, but remain edible, in large part because of the unique cool, airless, antiseptic and acidic environment of a peat bog. Firkins of such buried butter are a common archaeological find in Ireland; the National Museum of Ireland – Archaeology has some containing "a grayish cheese-like substance, partially hardened, not much like butter, and quite free from putrefaction."

The practice was most common in Ireland in the 11th–14th centuries; it ended entirely before the 19th century.

Industrialization

Like Ireland, France became well known for its butter, particularly in Normandy and Brittany. By the 1860s, butter had become so in demand in France that Emperor Napoleon III offered prize money for an inexpensive substitute to supplement France's inadequate butter supplies. A French chemist claimed the prize with the invention of margarine in 1869. The first margarine was beef tallow flavored with milk and worked like butter; vegetable margarine followed after the development of hydrogenated oils around 1900.

Gustaf de Laval's centrifugal cream separator sped up the butter-making process.

Until the 19th century, the vast majority of butter was made by hand, on farms. The first butter factories appeared in the United States in the early 1860s, after the successful introduction of cheese factories a decade earlier. In the late 1870s, the centrifugal cream separator was introduced, marketed most successfully by Swedish engineer Carl Gustaf Patrik de Laval. This dramatically sped up the butter-making process by eliminating the slow step of letting cream naturally rise to the top of milk. Initially, whole milk was shipped to the butter factories, and the cream separation took place there. Soon, though, cream-separation technology became small and inexpensive enough to introduce an additional efficiency: the separation was accomplished on the farm, and the cream alone shipped to the factory. By 1900, more than half the butter produced in the United States was factory made; Europe followed suit shortly after.

In 1920, Otto Hunziker authored *The Butter Industry, Prepared for Factory, School and Laboratory*, a well-known text in the industry that enjoyed at least three editions (1920, 1927, 1940). As part of the efforts of the American Dairy Science Association, Professor Hunziker and others published articles regarding: causes of tallowiness (an odor defect, distinct from rancidity, a taste defect); mottles (an aesthetic issue related to uneven color); introduced salts; the impact of creamery metals and liquids; and acidity measurement. These and other ADSA publications helped standardize practices internationally.

Butter also provided extra income to farm families. They used wood presses with carved decoration to press butter into pucks or small bricks to sell at nearby markets or general stores. The decoration identified the farm that produced the butter. This practice continued until production was mechanized and butter was produced in less decorative stick form. Today, butter presses remain in use for decorative purposes.

Per capita butter consumption declined in most western nations during the 20th century, in large part because of the rising popularity of margarine, which is less expensive and, until recent years, was perceived as being healthier. In the United States, margarine consumption overtook butter during the 1950s, and it is still the case today that more margarine than butter is eaten in the U.S. and the EU.

Size and Shape of Butter Packaging

Butter has traditionally been made into small, rectangular blocks by means of a pair of wooden butter paddles.

In the United States, butter is usually produced in 4-ounce sticks, wrapped in waxed or foiled paper and sold four to a one-pound carton. This practice is believed to have originated in 1907, when Swift and Company began packaging butter in this manner for mass distribution.

Western-pack shape butter

Due to historical differences in butter printers (machines that cut and package butter), these sticks are commonly produced in two different shapes:

- The dominant shape east of the Rocky Mountains is the Elgin, or Eastern-pack shape, named for a dairy in Elgin, Illinois. The sticks are 121 millimetres (4.8 in) long and 32 millimetres (1.3 in) wide and are typically sold stacked two by two in elongated cube-shaped boxes.

- West of the Rocky Mountains, butter printers standardized on a different shape that is now referred to as the Western-pack shape. These butter sticks are 80 millimetres (3.1 in) long and 38 millimetres (1.5 in) wide and are usually sold with four sticks packed side-by-side in a flat, rectangular box.

Both sticks contain the same amount of butter, although most butter dishes are designed for Elgin-style butter sticks.

The stick's wrapper is usually marked off as eight tablespoons (120 ml or 4.2 imp fl oz or 4.1 US fl oz); the actual volume of one stick is approximately nine tablespoons (130 ml or 4.6 imp fl oz or 4.4 US fl oz).

Outside of the United States, butter is packaged and sold by weight only, not by volume (fluid measure) nor by unit (stick), but the package shape remains approximately the same. The wrapper is usually a foil and waxed-paper laminate (the waxed paper is now a siliconised substitute, but is still referred to in some places as parchment, from the wrapping used in past centuries; and the term 'parchment-wrapped' is still employed where the paper alone is used, without the foil laminate).

In the UK and Ireland, and in some other regions historically accustomed to using British measures, this was traditionally ½lb and 1 lb packs; since metrication, pack sizes have changed to similar metric sizes such as 250g and 500g. In cooking (recipes), butter is specified and measured by weight only (grams or ounces); although melted butter could be measured by fluid measure (centiliters or fluid ounces), this is rare.

In the remainder of the metricated world, butter is packed and sold in 250g and 500g packs (roughly equivalent to the ½lb and 1 lb measures) and measured for cooking in grams or kilograms.

Butter for commercial and industrial use is packaged in plastic buckets, tubs, or drums, in quantities and units suited to the local market.

Worldwide

Butter Market, Lhasa, Tibet. 1993

In 1997, India produced 1,470,000 metric tons (1,620,000 short tons) of butter, most of which was consumed domestically. Second in production was the United States (522,000 t or 575,000 short tons), followed by France (466,000 t or 514,000 short tons), Germany (442,000 t or 487,000 short tons), and New Zealand (307,000 t or 338,000 short tons). France ranks first in per capita butter consumption with 8 kg per capita per year. In terms of absolute consumption, Germany was second after India, using 578,000 metric tons (637,000 short tons) of butter in 1997, followed by France (528,000 t or 582,000 short tons), Russia (514,000 t or 567,000 short tons), and the United States (505,000 t or 557,000 short tons). New Zealand, Australia, and the Ukraine are among the few nations that export a significant percentage of the butter they produce.

Different varieties are found around the world. *Smen* is a spiced Moroccan clarified butter, buried in the ground and aged for months or years. Yak butter is a speciality in Tibet; *tsampa*, barley flour mixed with yak butter, is a staple food. Butter tea is consumed in the Himalayan regions of Tibet, Bhutan, Nepal and India. It consists of tea served with intensely flavored—or "rancid"—yak butter and salt. In African and Asian developing nations, butter is traditionally made from sour milk rather than cream. It can take several hours of churning to produce workable butter grains from fermented milk.

Storage and Cooking

Hollandaise sauce served over white asparagus and potatoes.

Normal butter softens to a spreadable consistency around 15 °C (60 °F), well above refrigerator temperatures. The "butter compartment" found in many refrigerators may be one of the warmer sections inside, but it still leaves butter quite hard. Until recently, many refrigerators sold in New Zealand featured a "butter conditioner", a compartment kept warmer than the rest of the refrigerator—but still cooler than room temperature—with a small heater. Keeping butter tightly wrapped delays rancidity, which is hastened by exposure to light or air, and also helps prevent it from picking up other odors. Wrapped butter has a shelf life of several months at refrigerator temperatures. "French butter dishes" or "Acadian butter dishes" have a lid with a long interior lip, which sits in a container holding a small amount of water. Usually the dish holds just enough water to submerge the interior lip when the dish is closed. Butter is packed into the lid. The water acts as a seal to keep the butter fresh, and also keeps the butter from overheating in hot temperatures. This method lets butter sit on a countertop for several days without spoiling.

Once butter is softened, spices, herbs, or other flavoring agents can be mixed into it, producing what is called a *compound butter* or *composite butter* (sometimes also called *composed butter*). Compound butters can be used as spreads, or cooled, sliced, and placed onto hot food to melt into a sauce. Sweetened compound butters can be served with desserts; such hard sauces are often flavored with spirits.

When heated, butter quickly melts into a thin liquid.

Melted butter plays an important role in the preparation of sauces, most obviously in French cuisine. *Beurre noisette* (hazelnut butter) and *Beurre noir* (black butter) are sauces of melted butter cooked until the milk solids and sugars have turned golden or dark brown; they are often finished with an addition of vinegar or lemon juice. Hollandaise and béarnaise sauces are emulsions of egg yolk and melted butter; they are in essence mayonnaises made with butter instead of oil. Hollandaise and béarnaise sauces are stabilized with the powerful emulsifiers in the egg yolks, but butter itself contains enough emulsifiers—mostly remnants of the fat globule membranes—to form a stable emulsion on its own. *Beurre blanc* (white butter) is made by whisking butter into reduced vinegar or wine, forming an emulsion with the texture of thick cream. *Beurre monté* (prepared butter) is melted but still emulsified butter; it lends its name to the practice of "mounting" a sauce with butter: whisking cold butter into any water-based sauce at the end of cooking, giving the sauce a thicker body and a glossy shine—as well as a buttery taste.

In Poland, the butter lamb (*Baranek wielkanocny*) is a traditional addition to the Easter Meal for many Polish Catholics. Butter is shaped into a lamb either by hand or in a lamb-shaped mould. Butter is also used to make edible decorations to garnish other dishes.

Butter is used for sautéing and frying, although its milk solids brown and burn above 150 °C (250 °F)—a rather low temperature for most applications. The smoke point of butterfat is around 200 °C (400 °F), so clarified butter or ghee is better suited to frying.

Ghee has always been a common frying medium in India, where many avoid other animal fats for cultural or religious reasons.

Mixing melted butter with chocolate to make a brownie.

Butter fills several roles in baking, where it is used in a similar manner as other solid fats like lard, suet, or shortening, but has a flavor that may better complement sweet baked goods. Many cookie doughs and some cake batters are leavened, at least in part, by creaming butter and sugar together, which introduces air bubbles into the butter. The tiny bubbles locked within the butter expand in the heat of baking and aerate the cookie or cake. Some cookies like shortbread may have no other source of moisture but the water in the butter. Pastries like pie dough incorporate pieces of solid fat into the dough, which become flat layers of fat when the dough is rolled out. During baking, the fat melts away, leaving a flaky texture. Butter, because of its flavor, is a common choice for the fat in such a dough, but it can be more difficult to work with than shortening because of its low melting point. Pastry makers often chill all their ingredients and utensils while working with a butter dough.

Nutritional Information

As butter is essentially just the milk fat, it contains only traces of lactose, so moderate consumption of butter is not a problem for lactose intolerant people. People with milk allergies may still need to avoid butter, which contains enough of the allergy-causing proteins to cause reactions. Whole milk, butter and cream have high levels of saturated fat.

Butter is a good source of Vitamin A.

Comparative properties of common cooking fats (per 100 g)					
Type of fat	Total fat (g)	Saturated fat (g)	Monounsaturated fat (g)	Polyunsaturated fat (g)	Smoke point
Sunflower oil	100	11	20	69	225 °C (437 °F)

Sunflower oil (high oleic)	100	12	84	4	
Soybean oil	100	16	23	58	257 °C (495 °F)
Canola oil	100	7	63	28	205 °C (401 °F)
Olive oil	100	14	73	11	190 °C (374 °F)
Corn oil	100	15	30	55	230 °C (446 °F)
Peanut oil	100	17	46	32	225 °C (437 °F)
Rice bran oil	100	25	38	37	250 °C (482 °F)
Vegetable shortening (hydrogenated)	71	23	8	37	165 °C (329 °F)
Lard	100	39	45	11	190 °C (374 °F)
Suet	94	52	32	3	200 °C (392 °F)
Butter	81	51	21	3	150 °C (302 °F)
Coconut oil	100	86	6	2	177 °C (351 °F)

Churning (Butter)

Farmer girl churning butter

Churning is the process of shaking up cream (or whole milk) to make butter, and various forms of butter churn have been used for the purpose. In Europe from the Middle Ages until the Industrial Revolution, this was generally as simple as a barrel with a plunger in it, which was moved by hand. Afterward, mechanical means of churning were usually substituted and are considered to be far better.

Butter is essentially the fat of milk. It is usually made from sweet cream. In the USA, Ireland, the UK and the Nordic countries, salt is usually added to it. Unsalted (sweet) butters are most commonly used in the rest of Europe. However, it can also be made from acidulated or bacteriologically soured cream. Well into the 19th century butter was still made from cream that had been allowed to stand and sour naturally. The cream was then skimmed from the top of the milk and poured into a wooden tub.

Buttermaking was done by hand in butter churns. The natural souring process is, however, a very sensitive one and infection by foreign microorganisms often spoiled the result. Today's commercial buttermaking is a product of the knowledge and experience gained over the years in such matters as hygiene, bacterial acidifying and heat treatment, as well as the rapid technical development that has led to the advanced machinery now used. The commercial cream separator was introduced at the end of the 19th century, the continuous churn had been commercialized by the middle of the 20th century.

The Creation Process

Butter churner pot

Changing whole milk to butter is a process of transforming a fat-in-water emulsion (milk) to a water-in-fat emulsion (butter). Whole milk is a dilute emulsion of tiny fat globules surrounded by lipoprotein membranes that keep the fat globules separate from one another.

Butter is made from cream that has been separated from whole milk and then cooled; fat droplets clump more easily when hard rather than soft. However, making good butter also depends upon other factors, such as the fat content of the cream and its acidity.

A barrel-type butter churn.

The process can be summarized in 3 steps:

1. Churning physically agitates the cream until it ruptures the fragile membranes surrounding the milk fat. Once broken, the fat droplets can join with each other and form clumps of fat.

2. As churning continues, larger clusters of fat collect until they begin to form a network with the air bubbles that are generated by the churning; this traps the liquid and produces a foam. As the fat clumps increase in size, there are also fewer to enclose the air cells. So the bubbles pop, run together, and the foam begins to leak. This leakage is called buttermilk.

3. The cream separates into butter and buttermilk. The buttermilk is drained off, and the remaining butter is kneaded to form a network of fat crystals that becomes the continuous phase, or dispersion medium, of a water-in-fat emulsion. Working the butter also creates its desired smoothness. Eventually, the water droplets become so finely dispersed in the fat that butter's texture seems dry. Then it is frozen into cubes, then melted, then frozen again into bigger chunks to sell.

Types of Butter Churns

Butter churns have varied over time as technology and materials have changed.

1. Butter was first made by placing the cream in a container made from animal material and shaking until the milk has broken down into butter. Later wood, glass, ceramic or metal containers were used.

2. The first butter churns used a wooden container and a plunger to agitate the cream until butter formed.

3. Later butter churns used a container made from wood, ceramics or galvanized (zinc coated) iron that contained paddles. The hand-turned paddles were moved

through the cream quickly, breaking the cream up by mixing it with air. This allows the butter to be made faster than by simply agitating the cream.

4. Centrifugal cream separators allow the properties of centrifuge to be applied to butter making. Instead of having spinning paddles, the paddles are fixed and the container spins. This allows better separation of the butter from the buttermilk and water.

Churn with plunger
Ethnographic Museum of Western Liguria, Cervo, Italy

With electric mixers and food processors commonly available in most household kitchens, people can make butter in their own homes without a large churn. These small appliances are used to mix the cream until it is close to forming solid butter. This mixture is then mixed by hand to remove the buttermilk and water.

Historical Reference

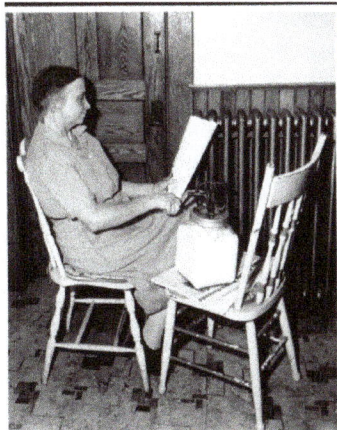
Churning butter (photo taken in 1944)

The *Household Cyclopedia* of 1881 instructs:

Let the cream be at the temperature of 55° to 60°, by a Fahrenheit thermometer; this is very important. If the weather be cold put boiling water into the churn for half an hour before you want to use it; when that is poured off strain in the cream through a butter cloth. When the butter is coming, which is easily ascertained by the sound, take off the lid, and with a small, flat board scrape down the sides of the churn, and do the same to the lid: this prevents waste. When the butter is come the butter-milk is to be poured off and spring water put into the churn, and turned for two or three minutes; this is to be then poured away and fresh added, and again the handle turned for a minute or two. Should there be the least milkiness when this is poured from the churn, more must be put in.

The butter is then to be placed on a board or marble slab and salted to taste; then with a cream cloth, wrung out in spring water, press all the moisture from it. When dry and firm make it up into rolls with flat boards. The whole process should be completed in three-quarters of an hour. In hot weather pains must be taken to keep the cream from reaching too high a heat. If the dairy be not cool enough, keep the cream-pot in the coldest water you can get; make the butter early in the morning, and place cold water in the churn for a while before it is used.

References

- McGee, Harold (2004). On Food and Cooking: The Science and Lore of the Kitchen. New York, New York: Scribner. ISBN 978-0-684-80001-1. LCCN 2004058999. OCLC 56590708.

- Soyer, Alexis (1977) [1853]. The Pantropheon or a History of Food and its Preparation in Ancient Times. Wisbech, Cambs.: Paddington Press. p. 172. ISBN 0-448-22976-5.

- The Culinary Institute of America (2011). The Professional Chef (9th ed.). Hoboken, New Jersey: John Wiley & Sons. ISBN 978-0-470-42135-2. OCLC 707248142.

- "What is sour cream. Sour cream for cooking recipes". Homecooking.about.com. 2010-06-14. Retrieved 2011-09-14.

Milk and Milk Processing

Milk is the primary source of nutrition; as an agricultural product it is extracted from animals such as cows, goats and camels. Homogenization, pasteurization, fouling, soured milk, raw milk and automatic milking are some of the aspects of milk and milk processing. The topics discussed in the section are of great importance to broaden the existing knowledge on milk.

Milk

Milk is a pale liquid produced by the mammary glands of mammals. It is the primary source of nutrition for infant mammals before they are able to digest other types of food. Early-lactation milk contains colostrum, which carries the mother's antibodies to its young and can reduce the risk of many diseases. It contains many other nutrients including protein and lactose.

As an agricultural product, milk is extracted from non-human mammals during or soon after pregnancy. Dairy farms produced about 730 million tonnes of milk in 2011, from 260 million dairy cows. India is the world's largest producer of milk, and is the leading exporter of skimmed milk powder, yet it exports few other milk products. The ever increasing rise in domestic demand for dairy products and a large demand-supply gap could lead to India being a net importer of dairy products in the future. The United States, India, China and Brazil are the world's largest exporters of milk and milk products. China and Russia were the world's largest importers of milk and milk products until 2016 when both countries became self-sufficient, contributing to a worldwide glut of milk.

Throughout the world, there are more than six billion consumers of milk and milk products. Over 750 million people live in dairy farming households.

Types of Consumption

There are two distinct types of milk consumption: a natural source of nutrition for all infant mammals and a food product for humans of all ages that is derived from other animals.

Nutrition for Infant Mammals

In almost all mammals, milk is fed to infants through breastfeeding, either directly or by expressing the milk to be stored and consumed later. The early milk from mammals is called colostrum. Colostrum contains antibodies that provide protection to the newborn baby as well as nutrients and growth factors. The makeup of the colostrum and the period of secretion varies from species to species.

A human baby feeding on their mother's milk

A goat kid feeding on their mother's milk

For humans, the World Health Organization recommends exclusive breastfeeding for six months and breastfeeding in addition to other food for at least two years. In some cultures it is common to breastfeed children for three to five years, and the period may be longer.

Fresh goats' milk is sometimes substituted for breast milk. This introduces the risk of the child developing electrolyte imbalances, metabolic acidosis, megaloblastic anemia, and a host of allergic reactions.

Food product for Humans

In many cultures of the world, especially the West, humans continue to consume milk beyond infancy, using the milk of other animals (especially cattle, goats and sheep) as a food product. Initially, the ability to digest milk was limited to children as adults did not produce lactase, an enzyme necessary for digesting the lactose in milk. Milk was therefore converted to curd, cheese and other products to reduce the levels of lactose. Thousands of years ago, a chance mutation spread in human populations in Europe that enabled the production of lactase in adulthood. This allowed milk to be used as a new source of nutrition which could sustain populations when other food sources failed. Milk is processed into a variety of dairy products such as cream, butter, yogurt, kefir, ice cream, and cheese. Modern industrial processes use milk to produce casein, whey protein, lactose, condensed milk, powdered milk, and many other food-additives and industrial products.

The Holstein Friesian cattle is the dominant breed in quintessential industrialized dairy farms today

Whole milk, butter and cream have high levels of saturated fat. The sugar lactose is found only in milk, forsythia flowers, and a few tropical shrubs. The enzyme needed to digest lactose, lactase, reaches its highest levels in the small intestine after birth and then begins a slow decline unless milk is consumed regularly. Those groups who do continue to tolerate milk, however, often have exercised great creativity in using the milk of domesticated ungulates, not only of cattle, but also sheep, goats, yaks, water buffalo, horses, reindeer and camels. The largest producer and consumer of cattle and buffalo milk in the world is India.

Per capita consumption of milk and milk products in selected countries in 2011			
Country	**Milk (liters)**	**Cheese (kg)**	**Butter (kg)**
Ireland	135.6	6.7	2.4
Finland	127.0	22.5	4.1
United Kingdom	105.9	10.9	3.0
Australia	105.3	11.7	4.0
Sweden	90.1	19.1	1.7
Canada	78.4	12.3	2.5
United States	75.8	15.1	2.8
Europe	62.8	17.1	3.6

🇧🇷 Brazil	55.7	3.6	0.4
🇫🇷 France	55.5	26.3	7.5
🇮🇹 Italy	54.2	21.8	2.3
🇩🇪 Germany	51.8	22.9	5.9
🇬🇷 Greece	49.1	23.4	0.7
🇳🇱 Netherlands	47.5	19.4	3.3
🇮🇳 India	39.5	-	3.5
🇨🇳 China	9.1	-	0.1

Terminology

The term *milk* is also used for white colored, non-animal beverages resembling milk in color and texture (milk substitutes) such as soy milk, rice milk, almond milk, and coconut milk. In addition, a substance secreted by pigeons to feed their young is called "crop milk" and bears some resemblance to mammalian milk, although it is not consumed as a milk substitute. Dairy relates to milk and milk production, e.g. dairy products. Milk can be synthesized in a laboratory, from water, fatty acids and proteins.

Evolution of Lactation

The mammary gland is thought to have derived from apocrine skin glands. It has been suggested that the original function of lactation (milk production) was keeping eggs moist. Much of the argument is based on monotremes (egg-laying mammals). The original adaptive significance of milk secretions may have been nutrition or immunological protection. This secretion gradually became more copious and accrued nutritional complexity over evolutionary time.

Tritylodontid cynodonts seem to have displayed lactation, based on their dental replacement patterns.

History

Humans first learned to regularly consume the milk of other mammals following the domestication of animals during the Neolithic Revolution or the development of agriculture. This development occurred independently in several places around the world from as early as 9000–7000 BC in Southwest Asia to 3500–3000 BC in the Americas. The most important dairy animals—cattle, sheep and goats—were first domesticated in Southwest Asia, although domestic cattle had been independently derived from wild aurochs populations several times since. Initially animals were kept for meat, and archaeologist Andrew Sherratt has suggested that dairying, along with the exploitation of domestic animals for hair and labor, began much later in a separate secondary products revolution in the fourth millennium BC. Sherratt's model is not supported by recent findings, based on the analysis of lipid residue in prehistoric pottery, that shows that dairying was practiced in the early phases of agriculture in Southwest Asia, by at least the seventh millennium BC.

Drinking milk in Germany in 1932

From Southwest Asia domestic dairy animals spread to Europe (beginning around 7000 BC but not reaching Britain and Scandinavia until after 4000 BC), and South Asia (7000–5500 BC). The first farmers in central Europe and Britain milked their animals. Pastoral and pastoral nomadic economies, which rely predominantly or exclusively on domestic animals and their products rather than crop farming, were developed as European farmers moved into the Pontic-Caspian steppe in the fourth millennium BC, and subsequently spread across much of the Eurasian steppe. Sheep and goats were introduced to Africa from Southwest Asia, but African cattle may have been independently domesticated around 7000–6000 BC. Camels, domesticated in central Arabia in the fourth millennium BC, have also been used as dairy animals in North Africa and the Arabian Peninsula. The earliest Egyptian records of burn treatments describe burn dressings using milk from mothers of male babies. In the rest of the world (i.e., East and Southeast Asia, the Americas and Australia) milk and dairy products were historically not a large part of the diet, either because they remained populated by hunter-gatherers who did not keep animals or the local agricultural economies did not include domesticated dairy species. Milk consumption became common in these regions comparatively recently, as a consequence of European colonialism and political domination over much of the world in the last 500 years.

In the Middle Ages, milk was called the "virtuous white liquor" because alcoholic beverages were more safe to consume than water.

Industrialization

Preserved Express Dairies three-axle milk tank wagon at the Didcot Railway Centre, based on an SR chassis

The growth in urban population coupled with the expansion of the railway network in the mid-19th century, brought about a revolution in milk production and supply. Individual railway firms began transporting milk from rural areas to London from the 1840s and 1850s. Possibly the first such instance was in 1846, when St Thomas's Hospital in Southwark contracted with milk suppliers outside London to provide milk by rail. The Great Western Railway was an early and enthusiastic adopter, and began to transport milk into London from Maidenhead in 1860, despite much criticism. By 1900, the company was transporting over 25 million gallons annually. The milk trade grew slowly through the 1860s, but went through a period of extensive, structural change in the 1870s and 1880s.

Milk transportation in Salem, Tamil Nadu

Urban demand began to grow, as consumer purchasing power increased and milk became regarded as a required daily commodity. Over the last three decades of the 19th century, demand for milk in most parts of the country doubled, or in some cases, tripled. Legislation in 1875 made the adulteration of milk illegal - this combined with a marketing campaign to change the image of milk. The proportion of rural imports by rail as a percentage of total milk consumption in London grew from under 5% in the

1860s to over 96% by the early 20th century. By that point, the supply system for milk was the most highly organized and integrated of any food product.

1959 milk supply in Oberlech, Vorarlberg, Austria

The first glass bottle packaging for milk was used in the 1870s. The first company to do so may have been the New York Dairy Company in 1877. The Express Dairy Company in England began glass bottle production in 1880. In 1884, Hervey Thatcher, an American inventor from New York, invented a glass milk bottle, called 'Thatcher's Common Sense Milk Jar', which was sealed with a waxed paper disk. Later, in 1932, plastic-coated paper milk cartons were introduced commercially.

In 1863, French chemist and biologist Louis Pasteur invented pasteurization, a method of killing harmful bacteria in beverages and food products. He developed this method while on summer vacation in Arbois, to remedy the frequent acidity of the local wines. He found out experimentally that it is sufficient to heat a young wine to only about 50−60 °C (122−140 °F) for a brief time to kill the microbes, and that the wine could be nevertheless properly aged without sacrificing the final quality. In honor of Pasteur, the process became known as "pasteurization". Pasteurization was originally used as a way of preventing wine and beer from souring. Commercial pasteurizing equipment was produced in Germany in the 1880s, and the process had been adopted in Copenhagen and Stockholm by 1885.

Overproduction

Continued improvements in the efficiency for the production of milk led to a worldwide glut of milk by 2016. Russia and China became self-sufficient and stopped importing milk. Canada has tried to restrict milk production by forcing new farmers/increased capacity to "buy in" at CN$24,000 per cow. Importing milk is prohibited. The European Union theoretically stopped subsidizing dairy farming in 2015. Direct subsidies were replaced by "environmental incentives" which results in the government buying milk when the price falls to €200 per 1,000 litres (220 imp gal; 260 US gal). The United States has a voluntary insurance program that pays farmers depending upon the price of milk and the cost of feed.

Sources of Milk

Modern dairy farm in Norway

The females of all mammal species can by definition produce milk, but cow's milk dominates commercial production. In 2011, FAO estimates 85% of all milk worldwide was produced from cows.

Human milk is not produced or distributed industrially or commercially; however, human milk banks collect donated human breastmilk and redistribute it to infants who may benefit from human milk for various reasons (premature neonates, babies with allergies, metabolic diseases, etc.) but who cannot breastfeed.

In the Western world, cow's milk is produced on an industrial scale and is by far the most commonly consumed form of milk. Commercial dairy farming using automated milking equipment produces the vast majority of milk in developed countries. Dairy cattle such as the Holstein have been bred selectively for increased milk production. About 90% of the dairy cows in the United States and 85% in Great Britain are Holsteins. Other dairy cows in the United States include Ayrshire, Brown Swiss, Guernsey, Jersey and Milking Shorthorn (Dairy Shorthorn).

Sources Aside from Cows

Other Significant Sources of Milk

Goats (2% of world's milk)

Buffaloes (11%)

Aside from cattle, many kinds of livestock provide milk used by humans for dairy products. These animals include buffalo, goat, sheep, camel, donkey, horse, reindeer and yak. The first four respectively produced about 11%, 2%, 1.4% and 0.2% of all milk worldwide in 2011.

In Russia and Sweden, small moose dairies also exist.

According to the US National Bison Association, American bison (also called American buffalo) are not milked commercially; however, various sources report cows resulting from cross-breeding bison and domestic cattle are good milk producers, and have been used both during the European settlement of North America and during the development of commercial Beefalo in the 1970s and 1980s.

Production Worldwide

Top ten cow milk producers in 2013		
Rank	**Country**	**Production (metric tonnes)**
1	🇺🇸 United States	91,271,058
2	🇮🇳 India	60,600,000
3	🇨🇳 China	35,310,000
4	🇧🇷 Brazil	34,255,236
5	🇩🇪 Germany	31,122,000
6	🇷🇺 Russia	30,285,969
7	🇫🇷 France	23,714,357
8	🇳🇿 New Zealand	18,883,000
9	🇹🇷 Turkey	16,655,009
10	🇬🇧 United Kingdom	13,941,000

Top ten sheep milk producers in 2013		
Rank	Country	Production (metric tonnes)
1	China	1,540,000
2	Turkey	1,101,013
3	Greece	705,000
4	Syria	684,578
5	Romania	632,582
6	Spain	600,568
7	Sudan	540,000
8	Somalia	505,000
9	Iran	470,000
10	Italy	383,837

Top ten goat milk producers in 2013		
Rank	Country	Production (metric tonnes)
1	India	5,000,000
2	Bangladesh	2,616,000
3	Sudan	1,532,000
4	Pakistan	801,000
5	Mali	720,000
6	France	580,694
7	Spain	471,999
8	Turkey	415,743
9	Somalia	400,000
10	Greece	340,000

Top ten buffalo milk producers in 2013		
Rank	Country	Production (metric tonnes)
1	India	70,000,000
2	Pakistan	24,370,000
3	China	3,050,000

4	Egypt	2,614,500
5	Nepal	1,188,433
6	Myanmar	309,000
7	Italy	194,893
8	Sri Lanka	65,000
9	Iran	65,000
10	Turkey	51,947

In 2012, the largest producer of milk and milk products was India followed by the United States of America, China, Pakistan and Brazil. All 28 European Union members together produced 153.8 million tonnes of milk in 2013, the largest by any politico-economic union.

Increasing affluence in developing countries, as well as increased promotion of milk and milk products, has led to a rise in milk consumption in developing countries in recent years. In turn, the opportunities presented by these growing markets have attracted investments by multinational dairy firms. Nevertheless, in many countries production remains on a small scale and presents significant opportunities for diversification of income sources by small farms. Local milk collection centers, where milk is collected and chilled prior to being transferred to urban dairies, are a good example of where farmers have been able to work on a cooperative basis, particularly in countries such as India.

Production Yields

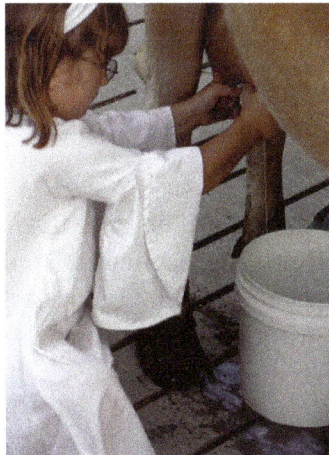

Child milking a cow by hand

FAO reports Israel dairy farms are the most productive in the world, with a yield of 12,546 kilograms (27,659 lb) milk per cow per year. This survey over 2001 and 2007 was conducted by ICAR (International Committee for Animal Recording) across 17 developed countries. The survey found that the average herd size in these developed countries increased from 74 to 99 cows per herd between 2001 to 2007. A dairy farm had an average of 19 cows per herd in Norway, and 337 in New Zealand. Annual milk production in the same period increased from 7,726 to 8,550 kg (17,033 to 18,850 lb) per cow in these developed countries. The lowest average production was in New Zealand at 3,974 kg (8,761 lb) per cow. The milk yield per cow depended on production systems, nutrition of the cows, and only to a minor extent different genetic potential of the animals. What the cow ate made the most impact on the production obtained. New Zealand cows with the lowest yield per year grazed all year, in contrast to Israel with the highest yield where the cows ate in barns with an energy-rich mixed diet.

The milk yield per cow in the United States, the world's largest cow milk producer, was 9,954 kg (21,945 lb) per year in 2010. In contrast, the milk yields per cow in India and China – the second and third largest producers – were respectively 1,154 kg (2,544 lb) and 2,282 kg (5,031 lb) per year.

Price

It was reported in 2007 that with increased worldwide prosperity and the competition of bio-fuel production for feed stocks, both the demand for and the price of milk had substantially increased worldwide. Particularly notable was the rapid increase of consumption of milk in China and the rise of the price of milk in the United States above the government subsidized price. In 2010 the Department of Agriculture predicted farmers would receive an average of $1.35 per US gallon of cow's milk (35 cents per liter), which is down 30 cents per gallon from 2007 and below the break-even point for many cattle farmers.

Grading

In the United States, there are two grades of milk, with grade A primarily used for direct sales and consumption in stores, and grade B used for indirect consumption, such as in cheese making or other processing.

The differences between the two grades are defined in the Wisconsin administrative code for Agriculture, Trade, and Consumer Protection, chapter 60. Grade B generally refers to milk that is cooled in milk cans, which are immersed in a bath of cold flowing water that typically is drawn up from an underground water well rather than using mechanical refrigeration.

Physical and Chemical Properties of Milk

Milk is an emulsion or colloid of butterfat globules within a water-based fluid that contains dissolved carbohydrates and protein aggregates with minerals. Because it is produced as a food source for the young, all of its contents provide benefits for growth. The principal requirements are energy (lipids, lactose, and protein), biosynthesis of non-essential amino acids supplied by proteins (essential amino acids and amino groups), essential fatty acids, vitamins and inorganic elements, and water.

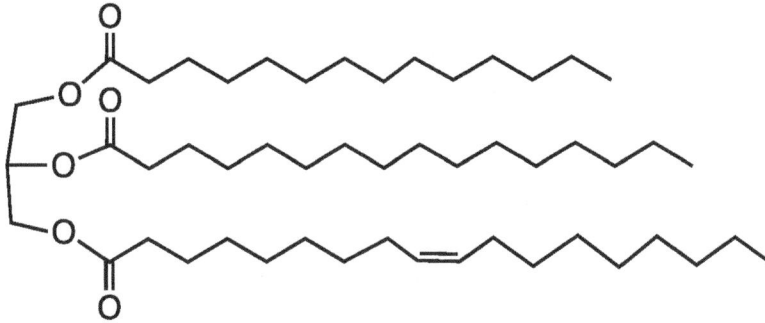

Butterfat is a triglyceride (fat) formed from fatty acids such as myristic, palmitic, and oleic acids.

pH

The pH of milk ranges from 6.4 to 6.8 and it changes over time. Milk from other bovines and non-bovine mammals varies in composition, but has a similar pH.

Lipids

Initially milk fat is secreted in the form of a fat globule surrounded by a membrane. Each fat globule is composed almost entirely of triacylglycerols and is surrounded by a membrane consisting of complex lipids such as phospholipids, along with proteins. These act as emulsifiers which keep the individual globules from coalescing and protect the contents of these globules from various enzymes in the fluid portion of the milk. Although 97–98% of lipids are triacylglycrols, small amounts of di- and monoacylglycerols, free cholesterol and cholesterol esters, free fatty acids, and phospholipids are also present. Unlike protein and carbohydrates, fat composition in milk varies widely in the composition due to genetic, lactational, and nutritional factor difference between different species.

Like composition, fat globules vary in size from less than 0.2 to about 15 micrometers in diameter between different species. Diameter may also vary between animals within a species and at different times within a milking of a single animal. In unhomogenized cow's milk, the fat globules have an average diameter of two to four micrometers and with homogenization, average around 0.4 micrometers. The fat-soluble vitamins A, D, E, and K along with essential fatty acids such as linoleic and linolenic acid are found within the milk fat portion of the milk.

Proteins

Normal bovine milk contains 30–35 grams of protein per liter of which about 80% is arranged in casein micelles. Total proteins in milk represent 3.2% of its composition (nutrition table).

Caseins

The largest structures in the fluid portion of the milk are "casein micelles": aggregates of several thousand protein molecules with superficial resemblance to a surfactant micelle, bonded with the help of nanometer-scale particles of calcium phosphate. Each casein micelle is roughly spherical and about a tenth of a micrometer across. There are four different types of casein proteins: $\alpha s1$-, $\alpha s2$-, β-, and κ-caseins. Collectively, they make up around 76–86% of the protein in milk, by weight. Most of the casein proteins are bound into the micelles. There are several competing theories regarding the precise structure of the micelles, but they share one important feature: the outermost layer consists of strands of one type of protein, k-casein, reaching out from the body of the micelle into the surrounding fluid. These kappa-casein molecules all have a negative electrical charge and therefore repel each other, keeping the micelles separated under normal conditions and in a stable colloidal suspension in the water-based surrounding fluid.

Milk contains dozens of other types of proteins beside caseins and including enzymes. These other proteins are more water-soluble than caseins and do not form larger structures. Because the proteins remain suspended in whey remaining when caseins coagulate into curds, they are collectively known as *whey proteins*. Whey proteins make up approximately 20% of the protein in milk by weight. Lactoglobulin is the most common whey protein by a large margin.

Salts, Minerals, and Vitamins

Minerals or milk salts, are traditional names for a variety of cations and anions within bovine milk. Calcium, phosphate, magnesium, sodium, potassium, citrate, and chlorine are all included as minerals and they typically occur at concentration of 5–40 mM. The milk salts strongly interact with casein, most notably calcium phosphate. It is present in excess and often, much greater excess of solubility of solid calcium phosphate. In addition to calcium, milk is a good source of many other vitamins. Vitamins A, B6, B12, C, D, K, E, thiamine, niacin, biotin, riboflavin, folates, and pantothenic acid are all present in milk.

Calcium Phosphate Structure

For many years the most accepted theory of the structure of a micelle was that it was composed of spherical casein aggregates, called submicelles, that were held together by calcium phosphate linkages. However, there are two recent models of the casein micelle that refute the distinct micellular structures within the micelle.

The first theory attributed to de Kruif and Holt, proposes that nanoclusters of calcium phosphate and the phosphopeptide fraction of beta-casein are the centerpiece to micellular structure. Specifically in this view, unstructured proteins organize around the calcium phosphate giving rise to their structure and thus no specific structure is formed.

The second theory proposed by Horne, the growth of calcium phosphate nanoclusters begins the process of micelle formation but is limited by binding phosphopeptide loop regions of the caseins. Once bound, protein-protein interactions are formed and polymerization occurs, in which K-casein is used as an end cap, to form micelles with trapped calcium phosphate nanoclusters.

Some sources indicate that the trapped calcium phosphate is in the form of Ca9(-PO4)6; whereas, others say it is similar to the structure of the mineral brushite CaH-PO4 -2H2O.

Carbohydrates and Miscellaneous Contents

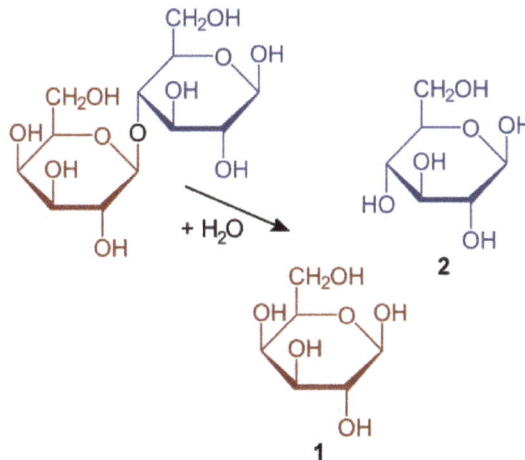

A simplified representation of a lactose molecule being broken down into glucose (2) and galactose (1)

Milk contains several different carbohydrate including lactose, glucose, galactose, and other oligosaccharides. The lactose gives milk its sweet taste and contributes approximately 40% of whole cow's milk's calories. Lactose is a disaccharide composite of two simple sugars, glucose and galactose. Bovine milk averages 4.8% anhydrous lactose, which amounts to about 50% of the total solids of skimmed milk. Levels of lactose are dependent upon the type of milk as other carbohydrates can be present at higher concentrations that lactose in milks.

Other components found in raw cow's milk are living white blood cells, mammary gland cells, various bacteria, and a large number of active enzymes.

Appearance

Both the fat globules and the smaller casein micelles, which are just large enough to deflect light, contribute to the opaque white color of milk. The fat globules contain some yellow-orange carotene, enough in some breeds (such as Guernsey and Jersey cattle) to impart a golden or "creamy" hue to a glass of milk. The riboflavin in the whey portion of milk has a greenish color, which sometimes can be discerned in skimmed milk or whey products. Fat-free skimmed milk has only the casein micelles to scatter light, and they tend to scatter shorter-wavelength blue light more than they do red, giving skimmed milk a bluish tint.

Processing

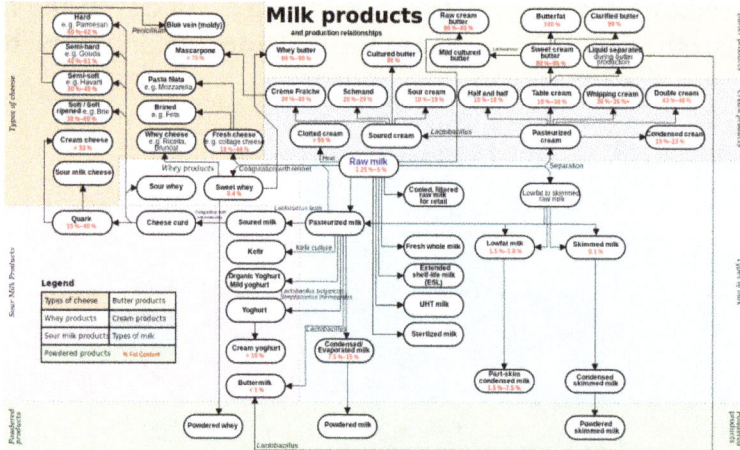

Milk products and productions relationships

In most Western countries, centralized dairy facilities process milk and products obtained from milk, such as cream, butter, and cheese. In the US, these dairies usually are local companies, while in the Southern Hemisphere facilities may be run by very large nationwide or trans-national corporations such as Fonterra.

Pasteurization

Pasteurization is used to kill harmful microorganisms by heating the milk for a short time and then immediately cooling it. Types of pasteurized milk include full cream, reduced fat, skim milk, calcium enriched, flavoured, and UHT. The standard high temperature short time (HTST) process produces a 99.999% reduction in the number of bacteria in milk, rendering it safe to drink for up to three weeks if continually refrigerated. Dairies print expiration dates on each container, after which stores remove any unsold milk from their shelves.

A side effect of the heating of pasteurization is that some vitamin and mineral content is lost. Soluble calcium and phosphorus decrease by 5%, thiamin and vitamin B12 by 10%, and vitamin C by 20%. Because losses are small in comparison to the large amount of the two B-vitamins present, milk continues to provide significant amounts of thiamin and vitamin B12. The loss of vitamin C is not nutritionally significant, as milk is not an important dietary source of vitamin C.

Microfiltration

Microfiltration is a process that partially replaces pasteurization and produces milk with fewer microorganisms and longer shelf life without a change in the taste of the milk. In this process, cream is separated from the whey and is pasteurized in the usual way, but the whey is forced through ceramic microfilters that trap 99.9% of microorganisms in the milk (as compared to 99.999% killing of microorganisms in standard HTST pasteurization). The whey then is recombined with the pasteurized cream to reconstitute the original milk composition.

Creaming and Homogenization

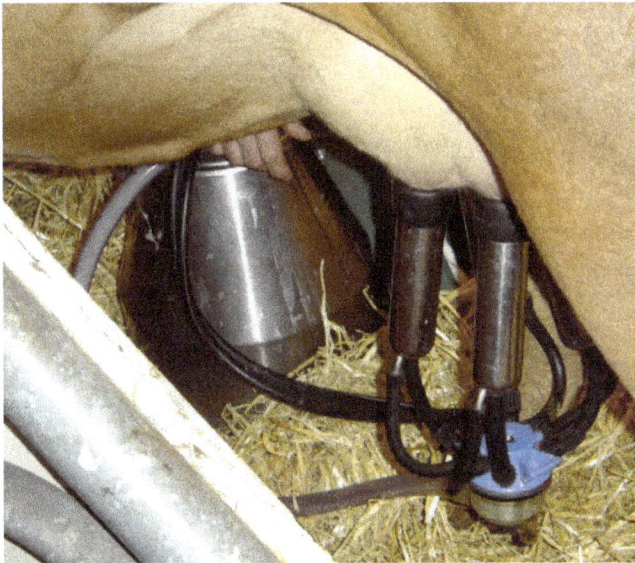
A milking machine in action

Upon standing for 12 to 24 hours, fresh milk has a tendency to separate into a high-fat cream layer on top of a larger, low-fat milk layer. The cream often is sold as a separate product with its own uses. Today the separation of the cream from the milk usually is accomplished rapidly in centrifugal cream separators. The fat globules rise to the top of a container of milk because fat is less dense than water. The smaller the globules, the more other molecular-level forces prevent this from happening. In fact, the cream rises in cow's milk much more quickly than a simple model would predict: rather than isolated globules, the fat in the milk tends to form into clusters containing about a million globules, held together by a number of minor whey proteins. These clusters rise faster than individual globules can. The fat globules in milk from goats, sheep, and water buffalo do not form clusters as readily and are smaller to begin with, resulting in a slower separation of cream from these milks.

Milk often is homogenized, a treatment that prevents a cream layer from separating out of the milk. The milk is pumped at high pressures through very narrow tubes, breaking up the fat globules through turbulence and cavitation. A greater number of smaller particles possess more total surface area than a smaller number of larger ones, and the original fat globule membranes cannot completely cover them. Casein micelles are attracted to the newly exposed fat surfaces. Nearly one-third of the micelles in the milk end up participating in this new membrane structure. The casein weighs down the globules and interferes with the clustering that accelerated separation. The exposed fat globules are vulnerable to certain enzymes present in milk, which could break down the fats and produce rancid flavors. To prevent this, the enzymes are inactivated by pasteurizing the milk immediately before or during homogenization.

Homogenized milk tastes blander but feels creamier in the mouth than unhomogenized. It is whiter and more resistant to developing off flavors. Creamline (or cream-top) milk is unhomogenized. It may or may not have been pasteurized. Milk that has undergone high-pressure homogenization, sometimes labeled as "ultra-homogenized", has a longer shelf life than milk that has undergone ordinary homogenization at lower pressures.

UHT

Ultra Heat Treatment (UHT), is a type of milk processing where the main aim is to destroy all bacteria in order to extend it's shelf life for up to 6 months once opened. Milk is firstly homogenized and then is heated to 138 Degrees Celsius for 1-3 seconds. The milk is then immediately cooled down and packed into a sterile container. As a result of this treatment, all the pathogenic bacteria within the milk is destroyed unlike when the milk is pasteurised. The milk will now keep for up for 6 months once unopened and refrigerated. But in this process there is a loss of vitamin B1 and vitamin C and there is also a slight change in the taste of the milk.

Nutrition and Health

The composition of milk differs widely among species. Factors such as the type of protein; the proportion of protein, fat, and sugar; the levels of various vitamins and minerals; and the size of the butterfat globules, and the strength of the curd are among those that may vary. For example:

- Human milk contains, on average, 1.1% protein, 4.2% fat, 7.0% lactose (a sugar), and supplies 72 kcal of energy per 100 grams.

- Cow's milk contains, on average, 3.4% protein, 3.6% fat, and 4.6% lactose, 0.7% minerals and supplies 66 kcal of energy per 100 grams.

Donkey and horse milk have the lowest fat content, while the milk of seals and whales may contain more than 50% fat.

Milk composition analysis, per 100 grams					
Constituents	**Unit**	**Cow**	**Goat**	**Sheep**	**Water buffalo**
Water	g	87.8	88.9	83.0	81.1
Protein	g	3.2	3.1	5.4	4.5
Fat	g	3.9	3.5	6.0	8.0
----Saturated fatty acids	g	2.4	2.3	3.8	4.2
----Monounsaturated fatty acids	g	1.1	0.8	1.5	1.7
----Polyunsaturated fatty acids	g	0.1	0.1	0.3	0.2
Carbohydrate (i.e the sugar form of lactose)	g	4.8	4.4	5.1	4.9
Cholesterol	mg	14	10	11	8
Calcium	mg	120	100	170	195
Energy	kcal	66	60	95	110
	kJ	275	253	396	463

Cow's Milk

These compositions vary by breed, animal, and point in the lactation period.

Milk fat percentages	
Cow breed	**Approximate percentage**
Jersey	5.2
Zebu	4.7
Brown Swiss	4.0
Holstein-Friesian	3.6

The protein range for these four breeds is 3.3% to 3.9%, while the lactose range is 4.7% to 4.9%.

Milk fat percentages may be manipulated by dairy farmers' stock diet formulation strategies. Mastitis infection can cause fat levels to decline.

Nutritional Value

Nutrient contents in %DV of common foods (raw, uncooked) per 100 g

Food	Protein DV	Protein Q	Fiber DV	A	B1	B2	B3	B5	B6	B9	B12	Ch.	C	D	E	K	Ca	Fe	Mg	P	K	Na	Zn	Cu	Mn	Se
cooking Reduction %				10	30	20	25		25	35	0	0	30				10	15	20	10	20	5	10	25		
Corn	20	55	6	1	13	4	16	4	19	19	0	0	0	0	0	1	1	11	31	34	15	1	20	10	42	0
Rice	14	71	1.3	0	12	3	11	20	5	2	0	0	0	0	0	0	1	9	6	7	2	0	8	9	49	22
Wheat	27	51	40	0	28	7	34	19	21	11	0	0	0	0	0	0	3	20	36	51	12	0	28	28	151	128
Soybean	73	132	0	31	58	51	8	8	19	94	0	24	10	0	4	59	28	87	70	70	51	0	33	83	126	25
Pigeon pea	43	91	1	50	43	11	15	13	13	114	0	0	0	0	0	0	13	29	46	37	40	1	18	53	90	12
Potato	4	112	7.3	0	5	2	5	3	15	4	0	0	33	0	0	2	1	4	6	6	12	0	2	5	8	0
Sweet potato	3	82	10	284	5	4	3	8	10	3	0	0	4	0	1	2	3	3	6	5	10	2	2	8	13	1
Spinach	6	119	7.3	188	5	11	4	1	10	49	0	4.5	47	0	10	604	10	15	20	5	16	3	4	6	45	1
Dill	7	32	7	154	4	17	8	4	9	38	0	0	142	0	0	0	21	37	14	7	21	3	6	7	63	0
Carrots	2		9.3	334	4	3	5	3	7	5	0	0	10	0	3	16	3	2	3	4	9	3	2	2	7	0
Guava	5	24	18	12	4	2	5	5	6	12	0	0	381	0	4	3	2	1	5	4	12	0	2	11	8	1
Papaya	1	7	5.6	22	2	2	2	2	1	10	0	0	103	0	4	3	2	1	2	1	7	0	0	1	1	1
Pumpkin	2	56	1.6	184	3	6	3	3	3	4	0	0	15	0	5	1	2	4	3	4	10	0	2	6	6	0
Sunflower oil	0	0	0	0	0	0	0	0	0	0	0	0	0	0	205	7	0	0	0	0	0	0	0	0	0	0
Egg	25	136	0	10	5	28	0	14	7	12	22	45	0	9	5	0	5	10	3	19	4	6	7	5	2	45
Milk	6	138	0	2	3	11	1	4	2	1	7	2.6	0	0	0	0	11	0	2	9	4	2	3	1	0	5

Ch. = Choline; Ca = Calcium; Fe = Iron; Mg = Magnesium; P = Phosphorus; K = Potassium; Na = Sodium; Zn = Zinc; Cu = Copper; Mn = Manganese; Se = Selenium; %DV = % daily value i.e. % of DRI (Dietary Reference Intake) Note: All nutrient values including protein and fiber are in %DV per 100 grams of the food item. Significant values are highlighted in light Gray color and bold letters. Cooking reduction = % Maximum typical reduction in nutrients due to boiling without draining for ovo-lacto-vegetables group Q = Quality of Protein in terms of completeness without adjusting for digestability.

Processed cow's milk was formulated to contain differing amounts of fat during the 1950s. One cup (250 ml) of 2%-fat cow's milk contains 285 mg of calcium, which represents 22% to 29% of the daily recommended intake (DRI) of calcium for an adult. Depending on its age, milk contains 8 grams of protein, and a number of other nutrients (either naturally or through fortification) including:

- Biotin
- Iodine
- Magnesium
- Pantothenic acid
- Potassium
- Riboflavin
- Selenium
- Thiamine
- Vitamin A
- Vitamin B_{12}
- Vitamins D
- Vitamin K

The amount of calcium from milk that is absorbed by the human body is disputed. Calcium from dairy products has a greater bioavailability than calcium from certain vegetables, such as spinach, that contain high levels of calcium-chelating agents, but a similar or lesser bioavailability than calcium from low-oxalate vegetables such as kale, broccoli, or other vegetables in the *Brassica* genus.

Milk as a calcium source has been questioned in media, but scientific research is lacking to support the hypothesis of acidosis induced by milk. The hypothesis in question being that acidosis would lead to leeching of calcium storages in bones to neutralize pH levels (also known as acid-ash hypothesis). Research has found no link between metabolic acidosis and consumption of milk.

Recommended Consumption

The U.S. federal government document *Dietary Guidelines for Americans, 2010* recommends consumption of three glasses of fat-free or low-fat milk for adults and children 9 and older (less for younger children) per day. This recommendation is disputed by some health researchers who call for more study of the issue, given that there are other sources for calcium and vitamin D. The researchers also claim that the recommendations have been unduly influenced by the American dairy industry, and that whole milk may be better for health due to its increased ability to satiate hunger.

Medical Research

There is recent evidence suggesting consumption of milk is effective at promoting muscle growth. Some studies have suggested that conjugated linoleic acid, which can be found in dairy products, is an effective supplement for reducing body fat.

With regards to the claim of milk promoting stronger bones, there has been no association between milk consumption or excess calcium intake and a reduced risk of bone fractures.

In 2011, *The Journal of Bone and Mineral Research* published a meta-analysis examining whether milk consumption might protect against hip fracture in middle-aged and older adults. Studies could find no association between drinking milk and lower rates of fractures. In 2014, *JAMA Pediatrics* published a report after monitoring almost 100,000 men and women for more than two decades. Subjects were asked to report on how much milk they had consumed as teenagers, and were followed to see if there is any association with a reduced chance of hip fractures later in life, it found there was not any. A study published in *The BMJ* that followed more than 45,000 men and 61,000 women in Sweden age 39 and older had similar results. Milk consumption in adults was associated with no protection for men, and an increased risk of fractures in women. The risk of any bone fracture increased 16 percent in women who drank three or more glasses daily, and the risk of a broken hip increased 60 percent. It was also associated with an increased risk of death in both sexes.

Milk and dairy products have the potential for causing serious infection in newborn infants. Unpasteurized milk and cheeses can promote the growth of *Listeria* bacteria. *Listeria monocytogenes* can also cause serious infection in an infant and pregnant woman and can be transmitted to her infant in utero or after birth. The infection has the potential of seriously harming or even causing the death of a preterm infant, an infant of low or very low birth weight, or an infant with an immune system defect or a congenital defect of the immune system. The presence of this pathogen can sometimes be determined by the symptoms that appear as a gastrointestinal illness in the mother. The mother can also acquire infection from ingesting food that contains other animal products such as, hot dogs, delicatessen meats, and cheese.

Lactose Intolerance

Lactose, the disaccharide sugar component of all milk, must be cleaved in the small intestine by the enzyme lactase, in order for its constituents, galactose and glucose, to be absorbed. Lactose intolerance is a condition in which people have symptoms due to not enough of the enzyme lactase in the small intestines. Those affected vary in the amount of lactose they can tolerate before symptoms develop. These may include abdominal pain, bloating, diarrhea, gas, and nausea. Severity depends on the amount a person eats or drinks. Those affected are usually able to drink at least one cup of milk without developing significant symptoms, with greater amounts tolerated if drunk with a meal or throughout the day.

Lactose intolerance does not cause damage to the gastrointestinal tract. There are four types: primary, secondary, developmental, and congenital. Primary lactose intolerance is when the amount of lactase decline as people age. Secondary lactose intolerance is due to injury to the small intestine such as from infection, celiac disease, inflammatory bowel disease, or other diseases. Developmental lactose intolerance may occur in premature babies and usually improves over a short period of time. Congenital lactose intolerance is an extremely rare genetic disorder in which little or no lactase is made from birth. When lactose intolerance is due to secondary lactase deficiency, treatment of the underlying disease allows lactase activity to return to normal levels. Lactose intolerance is different from a milk allergy.

The number of people with lactose intolerance is unknown. Some human populations have developed lactase persistence, in which lactase production continues into adulthood probably as a response to the benefits of being able to digest milk from farm animals. The percentage of the population that has a decrease in lactase as they age is less than 10% in Northern Europe and as high as 95% in parts of Asia and Africa.

Possible Harms

Some studies suggest that milk consumption may increase the risk of suffering from certain health problems. Cow's milk allergy (CMA) is an immunologically mediated adverse reaction, rarely fatal, to one or more cow's milk proteins. Milk from any mammal contains amino acids and microRNA which influence the drinker's metabolism and growth; this "programming" is beneficial for milk's natural consumers, namely infants of the same species as the milk producer, but post-infancy and trans-species milk consumption affects the mTORC1 metabolic pathway and may promote diseases of civilization such as obesity and diabetes.

Milk contains casein, a substance that breaks down in the human stomach to produce casomorphin, an opioid peptide. In the early 1990s it was hypothesized that casomorphin can cause or aggravate autism spectrum disorders, and casein-free diets are widely promoted. Studies supporting these claims have had significant flaws, and the data are inadequate to guide autism treatment recommendations.

The most recent assessment by the World Cancer Research Fund and the American Institute for Cancer Research found that most individual epidemiological studies showed increased risk of prostate cancer with increased intake of milk or dairy products. "Meta-analysis of cohort data produced evidence of a clear dose-response relationship between advanced/aggressive cancer risk with milk intake, and between all prostate cancer risk and milk and dairy products." Possible mechanisms proposed included inhibition of the conversion of vitamin D to its active metabolite, 1,25- dihydroxy vitamin D3 by calcium (which some evidence suggests increases cell proliferation in the prostate), and elevation of levels of insulin-like growth factor-1 (IGF-1). Several sources suggest a correlation between high calcium intake from milk, in particular, and prostate cancer, consistent with a calcium/vitamin D based mechanism. Overall, the WCRF/AICR panel concluded that "The evidence is inconsistent from both cohort and case-control studies. There is limited evidence suggesting that milk and dairy products are a cause of prostate cancer."

Medical studies also have shown a possible link between milk consumption and the exacerbation of diseases such as Crohn's disease, Hirschsprung's disease–mimicking symptoms in babies with existing cow's milk allergies, and the aggravation of Behçet's disease.

Flavored Milk in US Schools

Milk must be offered at every meal if a United States school district wishes to get reimbursement from the federal government. A quarter of the largest school districts in the US offer rice or soy milk and almost 17% of all US school districts offer lactose-free milk. Seventy-one percent of the milk served in US school cafeterias is flavored, causing some school districts to propose a ban because flavored milk has added sugars. (Though some flavored milk products use artificial sweeteners instead.) The Boulder, Colorado, school district banned flavored milk in 2009 and instead installed a dispenser that keeps the milk colder.

Bovine Growth Hormone Supplementation

Since November 1993, recombinant bovine somatotropin (rbST), also called rBGH, has been sold to dairy farmers with FDA approval. Cows produce bovine growth hormone naturally, but some producers administer an additional recombinant version of BGH which is produced through genetically engineered E. coli to increase milk production. Bovine growth hormone also stimulates liver production of insulin-like growth factor 1 (IGF1). The US Food and Drug Administration, the National Institutes of Health and the World Health Organization have reported that both of these compounds are safe for human consumption at the amounts present.

On June 9, 2006, the largest milk processor in the world and the two largest supermarkets in the United States – Dean Foods, Wal-Mart, and Kroger – announced that they are "on a nationwide search for rBGH-free milk." Milk from cows given rBST may be sold in the United States, and the FDA stated that no significant difference has been shown between milk derived from rBST-treated and that from non-rBST-treated cows. Milk that advertises that it comes from cows not treated with rBST, is required to state this finding on its label.

Cows receiving rBGH supplements may more frequently contract an udder infection known as mastitis. Problems with mastitis have led to Canada, Australia, New Zealand,

and Japan banning milk from rBST treated cows. Mastitis, among other diseases, may be responsible for the fact that levels of white blood cells in milk vary naturally.

rBGH is also banned in the European Union.

Criticism

Vegans and some other vegetarians do not consume milk for reasons mostly related to animal rights and environmental concerns. They may object to features of dairy farming including the necessity of keeping dairy cows pregnant, the killing of almost all the male offspring of dairy cows (either by disposal soon after birth, for veal production, or for beef), the routine separation of mother and calf soon after birth, other perceived inhumane treatment of dairy cattle, and culling of cows after their productive lives.

Some have criticized the American government's promotion of milk consumption. Their main concern is the financial interest that the American government has taken in the dairy industry, promoting milk as the best source of calcium. All United States schools that are a part of the federally funded National School Lunch Act are required by the federal government to provide milk for all students. The Office of Dietary Supplements recommends that healthy adults between ages 19 and 50 get about 1,000 mg of calcium per day.

Milk production is also resource intensive. On a global weighted average, 250 ml of milk production uses 250 liters of fresh water, or 1 gallon of milk uses 1,000 gallons of fresh water.

Varieties and Brands

Glass milk bottle used for home delivery service in the UK.

Milk products are sold in a number of varieties based on types/degrees of

- additives (e.g., vitamins),
- age (e.g., cheddar),
- coagulation (e.g., cottage cheese),
- farming method (e.g., organic, grass-fed).
- fat content (e.g., half and half),
- fermentation (e.g., buttermilk),
- flavoring (e.g., chocolate and strawberry),
- homogenization (e.g., cream top),
- reduction or elimination of lactose,
- mammal (e.g., cow, goat, sheep),
- packaging (e.g., bottle),
- pasteurization (e.g., raw milk),
- water content (e.g., dry milk)

Milk preserved by the UHT process does not need to be refrigerated before opening and has a longer shelf life than milk in ordinary packaging. It is typically sold unrefrigerated in the UK, US, Europe, Latin America, and Australia.

Reduction or elimination of Lactose

Lactose-free milk can be produced by passing milk over lactase enzyme bound to an inert carrier. Once the molecule is cleaved, there are no lactose ill effects. Forms are available with reduced amounts of lactose (typically 30% of normal), and alternatively with nearly 0%. The only noticeable difference from regular milk is a slightly sweeter taste due to the generation of glucose by lactose cleavage. It does not, however, contain more glucose, and is nutritionally identical to regular milk.

Finland, where approximately 17% of the Finnish-speaking population has hypolactasia, has had "HYLA" (acronym for *hydrolysed lactose*) products available for many years. Lactose of low-lactose level cow's milk products, ranging from ice cream to cheese, is enzymatically hydrolysed into glucose and galactose. The ultra-pasteurization process, combined with aseptic packaging, ensures a long shelf life. In 2001, Valio launched a lactose-free milk drink that is not sweet like HYLA milk but has the fresh taste of ordinary milk. Valio patented the chromatographic separation method to remove lactose. Valio also markets these products in Sweden, Estonia, Belgium, and the United States, where the company says ultrafiltration is used.

In the UK, where an estimated 4.7% of the population are affected by lactose intolerance, Lactofree produces milk, cheese, and yogurt products that contain only 0.03% lactose.

To aid digestion in those with lactose intolerance, milk with added bacterial cultures such as *Lactobacillus acidophilus* ("acidophilus milk") and bifidobacteria ("a/B milk") is available in some areas. Another milk with *Lactococcus lactis* bacteria cultures ("cultured buttermilk") often is used in cooking to replace the traditional use of naturally soured milk, which has become rare due to the ubiquity of pasteurization, which also kills the naturally occurring Lactococcus bacteria.

Additives and Flavoring

In areas where the cattle (and often the people) live indoors, commercially sold milk commonly has vitamin D added to it to make up for lack of exposure to UVB radiation.

Reduced-fat milks often have added vitamin A palmitate to compensate for the loss of the vitamin during fat removal; in the United States this results in reduced fat milks having a higher vitamin A content than whole milk.

Milk often has flavoring added to it for better taste or as a means of improving sales. Chocolate milk has been sold for many years and has been followed more recently by strawberry milk and others. Some nutritionists have criticized flavored milk for adding sugar, usually in the form of high-fructose corn syrup, to the diets of children who are already commonly obese in the US.

Distribution

Returning reusable glass milk bottles, used for home delivery service in the UK

Due to the short shelf life of normal milk, it used to be delivered to households daily in many countries; however, improved refrigeration at home, changing food shopping patterns because of supermarkets, and the higher cost of home delivery mean that daily deliveries by a milkman are no longer available in most countries.

A glass bottle of non-homogenized, organic, local milk from the US state of California. American milk bottles are generally rectangular in shape

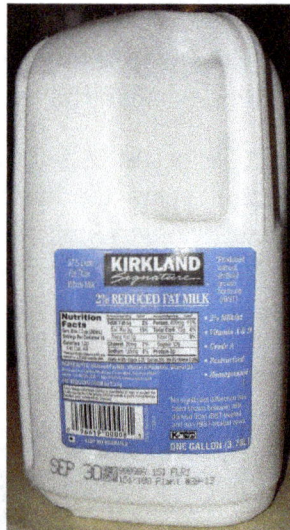

A rectangular milk jug design used by Costco and Sam's Club stores in the United States which allows for stacking and display of filled containers rather than being shipped to the store in milk crates and manual loading into a freezer display rack.

Australia and New Zealand

In Australia and New Zealand, prior to metrication, milk was generally distributed in 1 pint (568ml) glass bottles. In Australia and Ireland there was a government funded "free milk for school children" program, and milk was distributed at morning recess in 1/3 pint bottles. With the conversion to metric measures, the milk industry were concerned that the replacement of the pint bottles with 500ml bottles would result in a 13.6% drop in milk consumption; hence, all pint bottles were recalled and replaced by 600 mL bottles. With time, due to the steadily increasing cost of collecting, transporting, storing and cleaning glass bottles, they were replaced by cardboard cartons.

A number of designs were used, including a tetrahedron which could be close-packed without waste space, and could not be knocked over accidentally. (slogan: No more crying over spilt milk.) However, the industry eventually settled on a design similar to that used in the United States.

Milk is now available in a variety of sizes in cardboard cartons (250 mL, 375 mL, 600 mL, 1 liter and 1.5 liters) and plastic bottles (1, 2 and 3 liters). A significant addition to the marketplace has been "long-life" milk (UHT), generally available in 1 and 2 liter rectangular cardboard cartons. In urban and suburban areas where there is sufficient demand, home delivery is still available, though in suburban areas this is often 3 times per week rather than daily. Another significant and popular addition to the marketplace has been flavored milks – for example, as mentioned above, Farmers Union Iced Coffee outsells Coca-Cola in South Australia.

India

In rural India, milk is home delivered, daily, by local milkmen carrying bulk quantities in a metal container, usually on a bicycle. In other parts of metropolitan India, milk is usually bought or delivered in plastic bags or cartons via shops or supermarkets.

The current milk chain flow in India is from milk producer to milk collection agent. Then it is transported to a milk chilling center and bulk transported to the processing plant, then to the sales agent and finally to the consumer.

A 2011 survey by the Food Safety and Standards Authority of India found that nearly 70 per cent of samples had not conformed to the standards set for milk. The study found that due to lack of hygiene and sanitation in milk handling and packaging, detergents (used during cleaning operations) were not washed properly and found their way into the milk. About eight per cent of samples in the survey were found to have detergents, which are hazardous to health.

Pakistan

In Pakistan, milk is supplied in jugs. Milk has been a staple food, especially among the pastoral tribes in this country.

United Kingdom

Since the late 1990s, milk-buying patterns have changed drastically in the UK. The classic milkman, who travels his local milk round (route) using a milk float (often battery powered) during the early hours and delivers milk in 1 pint glass bottles with aluminium foil tops directly to households, has almost disappeared. Two of the main reasons for the decline of UK home deliveries by milkmen are household refrigerators (which lessen the need for daily milk deliveries) and private car usage (which has increased supermarket shopping). Another factor is that it is cheaper to purchase milk from a supermarket than from home delivery. In 1996, more than 2.5 billion liters of milk were still being delivered by milkmen, but by 2006 only 637 million liters (13% of milk consumed) was delivered by some 9,500 milkmen. By 2010, the estimated number of milkmen had dropped to 6,000. Assuming that delivery per milkman is the same as it was in 2006, this means milkmen deliveries now only account for 6–7% of all milk consumed by UK households (6.7 billion liters in 2008/2009).

Almost 95% of all milk in the UK is thus sold in shops today, most of it in plastic bottles of various sizes, but some also in milk cartons. Milk is hardly ever sold in glass bottles in UK shops.

United States

Getting milk at the back door ~ 1940

In the United States, glass milk bottles have been replaced mostly with milk cartons and plastic jugs. Gallons of milk are almost always sold in jugs, while half gallons and quarts may be found in both paper cartons and plastic jugs, and smaller sizes are almost always in cartons.

The "half pint" .5 US pints (0.24 l; 0.42 imp pt) milk carton is the traditional unit as a component of school lunches, though some companies have replaced that unit size with a plastic bottle, which is also available at retail in 6- and 12-pack size.

Packaging

Glass milk bottles are now rare. Most people purchase milk in bags, plastic bottles, or plastic-coated paper cartons. Ultraviolet (UV) light from fluorescent lighting can alter the flavor of milk, so many companies that once distributed milk in transparent or highly translucent containers are now using thicker materials that block the UV light. Milk comes in a variety of containers with local variants:

Australia and New Zealand

Distributed in a variety of sizes, most commonly in aseptic cartons for up to 1.5 liters, and plastic screw-top bottles beyond that with the following volumes; 1.1 L, 2 L, and 3 L. 1 liter milk bags are starting to appear in supermarkets, but have not yet proved popular. Most UHT-milk is packed in 1 or 2 liter paper containers with a sealed plastic spout.

Brazil

Used to be sold in cooled 1 liter bags, just like in South Africa. Today the most common form is 1 liter aseptic cartons containing UHT skimmed, semi-skimmed or whole milk, although the plastic bags are still in use for pasteurized milk. Higher grades of pasteurized milk can be found in cartons or plastic bottles. Sizes other than 1 liter are rare.

Canada

1.33 liter plastic bags (sold as 4 liters in 3 bags) are widely available in some areas (especially the Maritimes, Ontario and Quebec), although the 4 liter plastic jug has supplanted them in western Canada. Other common packaging sizes are 2 liter, 1 liter, 500 mL, and 250 mL cartons, as well as 4 liter, 1 liter, 250 mL aseptic cartons and 500 mL plastic jugs.

Chile

Distributed most commonly in aseptic cartons for up to 1 liter, but smaller, snack-sized cartons are also popular. The most common flavors, besides the natural presentation, are chocolate, strawberry and vanilla.

China

Sweetened milk is a drink popular with students of all ages and is often sold in small plastic bags complete with straw. Adults not wishing to drink at a banquet often drink milk served from cartons or milk tea.

Colombia

Sells milk in 1 liter plastic bags.

Croatia, Bosnia and Herzegovina, Serbia, Montenegro

UHT milk (*trajno mlijeko/trajno mleko*/трајно млеко) is sold in 500 mL and 1 L (sometimes also 200 ml) aseptic cartons. Non-UHT pasteurized milk (*svježe mlijeko/sveže mleko*/свеже млеко) is most commonly sold in 1 L and 1.5 L PET bottles, though in Serbia one can still find milk in plastic bags.

Estonia

Commonly sold in 1 L bags or 0.33 L, 0.5 L, 1 L or 1.5 L cartons.

Parts of Europe

Sizes of 500 mL, 1 liter (the most common), 1.5 liters, 2 liters and 3 liters are commonplace.

Finland

Commonly sold in 1 L or 1.5 L cartons, in some places also in 2 dl and 5 dl cartons.

Germany

Commonly sold in 1-liter cartons. Sale in 1-liter plastic bags (common in the 1980s) now rare.

Hong Kong

Milk is sold in glass bottles (220 mL), cartons (236 mL and 1 L), plastic jugs (2 liters) and aseptic cartons (250 mL).

India

Commonly sold in 500 mL plastic bags and in bottles in some parts like in west. It is still customary to serve the milk boiled, despite pasteurization. Milk is often buffalo milk. Flavored milk is sold in most convenience stores in waxed cardboard containers. Convenience stores also sell many varieties of milk (such as flavored and ultra-pasteurized) in different sizes, usually in aseptic cartons.

Indonesia

Usually sold in 1 liter cartons, but smaller, snack-sized cartons are available.

Israel

Non-UHT milk is most commonly sold in 1 liter waxed cardboard boxes and 1 liter plastic bags. It may also be found in 1.5 L and 2 L waxed cardboard boxes, 2 L plastic jugs and 1 L plastic bottles. UHT milk is available in 1 liter (and less commonly also in 0.5 L) carton "bricks".

Japan

Commonly sold in 1 liter waxed paperboard cartons. In most city centers there is also home delivery of milk in glass jugs. As seen in China, sweetened and flavored milk drinks are commonly seen in vending machines.

Kenya

Milk in Kenya is mostly sold in plastic-coated aseptic paper cartons supplied in 300 ml, 500 ml or 1 liter volumes. In rural areas, milk is stored in plastic bottles or gourds. The standard unit of measuring milk quantity in Kenya is a liter.

Pakistan

Milk is supplied in 500 ml plastic bags and carried in jugs from rural to cities for selling

Philippines

Milk is supplied in 1000 ml plastic bottles and delivered from factories to cities for selling.

Poland

UHT milk is mostly sold in aseptic cartons (500 mL, 1 L, 2 L), and non-UHT in 1 L plastic bags or plastic bottles. Milk, UHT is commonly boiled, despite being pasteurized.

South Africa

Commonly sold in 1 liter bags. The bag is then placed in a plastic jug and the corner cut off before the milk is poured.

South Korea

Sold in cartons (180 mL, 200 mL, 500 mL 900 mL, 1 L, 1.8 L, 2.3 L), plastic jugs (1 L and 1.8 L), aseptic cartons (180 mL and 200 mL) and plastic bags (1 L).

Sweden

Commonly sold in 0.3 L, 1 L or 1.5 L cartons and sometimes as plastic or glass milk bottles.

Turkey

Commonly sold in 500 mL or 1L cartons or special plastic bottles. UHT milk is more popular. Milkmen also serve in smaller towns and villages.

United Kingdom

Most stores stock imperial sizes: 1 pint (568 mL), 2 pints (1.136 L), 4 pints (2.273 L), 6 pints (3.408 L) or a combination including both metric and imperial sizes. Glass milk bottles delivered to the doorstep by the milkman are typically pint-sized and are returned empty by the householder for repeated reuse. Milk is sold at supermarkets in either aseptic cartons or HDPE bottles. Supermarkets have also now begun to introduce milk in bags, to be poured from a proprietary jug and nozzle.

United States

Commonly sold in gallon (3.78 L), half-gallon (1.89 L) and quart (0.94 L) containers of natural-colored HDPE resin, or, for sizes less than one gallon, cartons of waxed paperboard. Bottles made of opaque PET are also becoming common-place for smaller, particularly metric, sizes such as one liter. The US single-serving size is usually the half-pint (about 240 mL). Less frequently, dairies deliver milk directly to consumers, from coolers filled with glass bottles which are typically half-gallon sized and returned for reuse. Some convenience store chains in the United States (such as Kwik Trip in the Midwest) sell milk in half-gallon bags, while another rectangular cube gallon container design used for easy stacking in shipping and displaying is used by warehouse clubs such as Costco and Sam's Club, along with some Wal-Mart stores.

Uruguay

Commonly sold in 1 liter bags. The bag is then placed in a plastic jug and the corner cut off before the milk is poured.

Practically everywhere, condensed milk and evaporated milk are distributed in metal cans, 250 and 125 mL paper containers and 100 and 200 mL squeeze tubes, and powdered milk (skim and whole) is distributed in boxes or bags.

Spoilage and Fermented Milk Products

When raw milk is left standing for a while, it turns "sour". This is the result of fermentation, where lactic acid bacteria ferment the lactose in the milk into lactic acid. Prolonged fermentation may render the milk unpleasant to consume. This fermentation process is exploited by the introduction of bacterial cultures (e.g. *Lactobacilli sp., Streptococcus sp., Leuconostoc sp.*, etc.) to produce a variety of fermented milk products. The reduced pH from lactic acid accumulation denatures proteins and causes the milk to undergo a variety of different transformations in appearance and texture, ranging from an aggregate to smooth consistency. Some of these products include sour cream, yogurt, cheese, buttermilk, viili, kefir, and kumis.

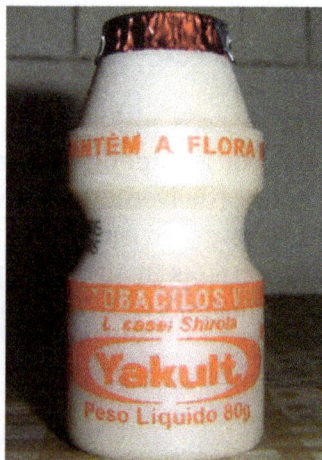

Yakult, a probiotic milk-like product made by fermenting a mixture of skimmed milk with a special strain of the bacterium *Lactobacillus casei Shirota*.

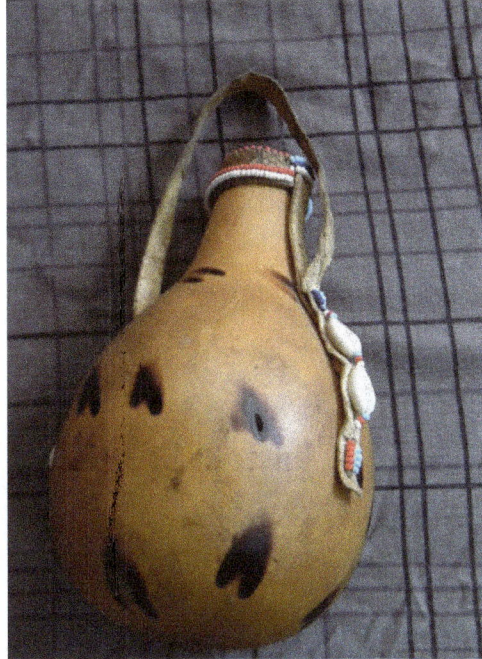

Gourd used by Kalenjins to prepare a local version of fermented milk called mursik.

Pasteurization of cow's milk initially destroys any potential pathogens and increases the shelf life, but eventually results in spoilage that makes it unsuitable for consumption. This causes it to assume an unpleasant odor, and the milk is deemed non-consumable due to unpleasant taste and an increased risk of food poisoning. In raw milk, the presence of lactic acid-producing bacteria, under suitable conditions, ferments the lactose present to lactic acid. The increasing acidity in turn prevents the growth of other organisms, or slows their growth significantly. During pasteurization, however, these lactic acid bacteria are mostly destroyed.

In order to prevent spoilage, milk can be kept refrigerated and stored between 1 and 4 °C (34 and 39 °F) in bulk tanks. Most milk is pasteurized by heating briefly and then refrigerated to allow transport from factory farms to local markets. The spoilage of milk can be forestalled by using ultra-high temperature (UHT) treatment. Milk so treated can be stored unrefrigerated for several months until opened but has a characteristic "cooked" taste. Condensed milk, made by removing most of the water, can be stored in cans for many years, unrefrigerated, as can evaporated milk. The most durable form of milk is powdered milk, which is produced from milk by removing almost all water. The moisture content is usually less than 5% in both drum- and spray-dried powdered milk.

Freezing of milk can cause fat globule aggregation upon thawing, resulting in milky layers and butterfat lumps. These can be dispersed again by warming and stirring the milk. It can change the taste by destruction of milk-fat globule membranes, releasing oxidized flavors.

Language and Culture

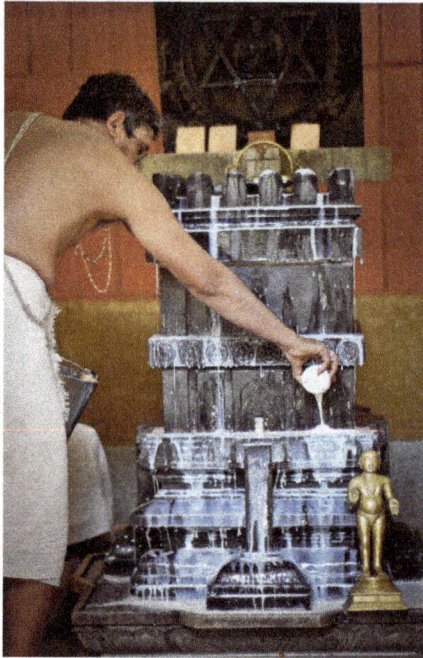

Hindu Abhisheka ritual in Agara, Bangalore Rural District, Karnataka.

The importance of milk in human culture is attested to by the numerous expressions embedded in our languages, for example, "the milk of human kindness". In ancient Greek mythology, the goddess Hera spilled her breast milk after refusing to feed Heracles, resulting in the Milky Way.

In many African and Asian countries, butter is traditionally made from fermented milk rather than cream. It can take several hours of churning to produce workable butter grains from fermented milk.

Holy books have also mentioned milk. The Bible contains references to the 'Land of Milk and Honey'. In the Qur'an, there is a request to wonder on milk as follows: 'And surely in the livestock there is a lesson for you, We give you to drink of that which is in their bellies from the midst of digested food and blood, pure milk palatable for the drinkers.'(16-The Honeybee, 66). The Ramadan fast is traditionally broken with a glass of milk and dates.

Abhisheka is conducted by Hindu and Jain priests, by pouring libations on the image of a deity being worshipped, amidst the chanting of mantras. Usually offerings such as milk, yogurt, ghee, honey may be poured among other offerings depending on the type of abhishekam being performed.

To milk someone, in the vernacular of many English-speaking countries, is to take advantage of the person.

The word "milk" has had many slang meanings over time. In the 19th century, milk was used to describe a cheap alcoholic drink made from methylated spirits mixed with water. The word was also used to mean defraud, to be idle, to intercept telegrams addressed to someone else, and a weakling or 'milksop'. In the mid-1930s, the word was used in Australia meaning to siphon gas from a car.

Other Uses

Besides serving as a beverage or source of food, milk has been described as used by farmers and gardeners as an organic fungicide and fertilizer, however, its effectiveness is debated. Diluted milk solutions have been demonstrated to provide an effective method of preventing powdery mildew on grape vines, while showing it is unlikely to harm the plant.

Homogenization (Chemistry)

Homogenization or homogenisation is any of several processes used to make a mixture of two mutually non-soluble liquids the same throughout. This is achieved by turning one of the liquids into a state consisting of extremely small particles distributed uniformly throughout the other liquid. A typical example is the homogenization of milk, where the milk fat globules are reduced in size and dispersed uniformly through the rest of the milk.

Definition

Homogenization is the process of converting two immiscible liquids (i.e. liquids that are not soluble, in all proportions, one in another) into an emulsion (an emulsion is a type of colloid, which is a substance microscopically dispersed throughout another substance; when both the dispersed and the continuous substances are liquids, the colloid is called an emulsion). Sometimes two types of homogenization are distinguished: primary homogenization, when the emulsion is created directly from separate liquids; and secondary homogenization, when the emulsion is created by the reduction in size of droplets in an existing emulsion. Homogenization is achieved by a mechanical device called a *homogenizer*.

Application

One of the oldest applications of homogenization is in milk processing. It is normally preceded by "standardization" (the mixing of several different milking herds and/or dairies to produce a more consistent raw milk prior to processing and to prevent, reduce and delay natural separation of cream from the rest of the emulsion). The fat in milk normally separates from the water and collects at the top. Homogenization breaks the fat into smaller sizes so it no longer separates, allowing the sale of non-separating milk at any fat specification.

Methods

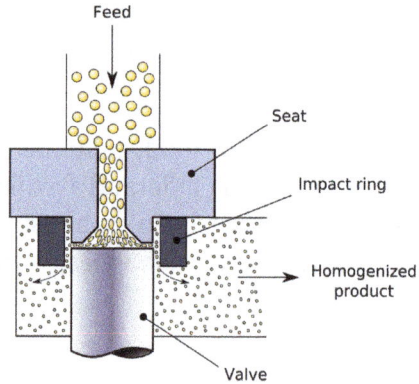

Homogenizing valve, a method to homogenize at high pressure.

Milk homogenization is accomplished by mixing massive amounts of harvested milk to create a constant, then forcing the milk at high pressure through small holes. Yet another method of homogenization uses extruders, hammermills, or colloid mills to mill (grind) solids. Milk homogenization is an essential tool of the milk food industry to prevent creating various levels of flavor and fat concentration.

Another application of homogenization is in soft drinks like cola products. The reactant mixture is rendered to intense homogenization, to as much as 35,000 psi, so that various constituents do not separate out during storage or distribution.

Pasteurization

Cream pasteurizing and cooling coils at Murgon Butter Factory, 1939

Pasteurization or pasteurisation is a process that kills microbes (mainly bacteria) in food and drink, such as milk, juice, canned food, and others.

It was invented by French scientist Louis Pasteur during the nineteenth century. In 1864 Pasteur discovered that heating beer and wine was enough to kill most of the bacteria that caused spoilage, preventing these beverages from turning sour. The process achieves this by eliminating pathogenic microbes and lowering microbial numbers to prolong the quality of the beverage. Today, pasteurisation is used widely in the dairy industry and other food processing industries to achieve food preservation and food safety.

Unlike sterilization, pasteurization is not intended to kill all microorganisms in the food. Instead, it aims to reduce the number of viable pathogens so they are unlikely to cause disease (assuming the pasteurized product is stored as indicated and is consumed before its expiration date). Commercial-scale sterilization of food is not common because it adversely affects the taste and quality of the product. Certain foods, such as dairy products, may be superheated to ensure pathogenic microbes are destroyed.

Pasteurization Conditions

Minimum pasteurization requirements for milk products are shown in Table 1 below, and are based on regulations outlined in the Grade A Pasteurized Milk Ordinance (PMO). These conditions were determined to be the minimum processing conditions needed to kill Coxiella burnetii, the organism that causes Q fever in humans, which is the most heat resistant pathogen currently recognized in milk. Milk can be pasteurized using processing times and temperatures greater than the required minimums.

Pasteurization can be done as a batch or a continuous process. A vat pasteurizer consists of a temperature-controlled, closed vat. The milk is pumped into the vat, the milk is heated to the appropriate temperature and held at that temperature for the appropriate time and then cooled. The cooled milk is then pumped out of the vat to the rest of the processing line, for example to the bottling station or cheese vat. Batch pasteurization is still used in some smaller processing plants. The most common process used for fluid milk is the continuous process. The milk is pumped from the raw milk silo to a holding tank that feeds into the continuous pasteurization system. The milk continuously flows from the tank through a series of thin plates that heat up the milk to the appropriate temperature. The milk flow system is set up to make sure that the milk stays at the pasteurization temperature for the appropriate time before it flows through the cooling area of the pasteurizer. The cooled milk then flows to the rest of the processing line, for example to the bottling station. There are several options for temperatures and times available for continuous processing of refrigerated fluid milk. Although processing conditions are defined for temperatures above 200 °F, they are rarely used because they can impart an undesirable cooked flavor to milk.

Alcoholic Beverages

The process of heating wine for preservation purposes has been known in China since 1117, and was documented in Japan in the diary *Tamonin-nikki*, written by a series of monks between 1478 and 1618.

Much later, in 1768, an Italian priest and scientist Lazzaro Spallanzani proved experimentally that heat killed bacteria, and that they do not reappear if the product is hermetically sealed. In 1795, a Parisian chef and confectioner named Nicolas Appert began experimenting with ways to preserve foodstuffs, succeeding with soups, vegetables, juices, dairy products, jellies, jams, and syrups. He placed the food in glass jars, sealed them with cork and sealing wax and placed them in boiling water. In that same year, the French military offered a cash prize of 12,000 francs for a new method to preserve food. After some 14 or 15 years of experimenting, Appert submitted his invention and won the prize in January 1810. Later that year, Appert published *L'Art de conserver les substances animales et végétales* (or *The Art of Preserving Animal and Vegetable Substances*). This was the first cookbook of its kind on modern food preservation methods.

La Maison Appert (English: The House of Appert), in the town of Massy, near Paris, became the first food-bottling factory in the world, preserving a variety of food in sealed bottles. Appert's method was to fill thick, large-mouthed glass bottles with produce of every description, ranging from beef and fowl to eggs, milk and prepared dishes. His greatest success for publicity was an entire sheep. He left air space at the top of the bottle, and the cork would then be sealed firmly in the jar by using a vise. The bottle was then wrapped in canvas to protect it, while it was dunked into boiling water and then boiled for as much time as Appert deemed appropriate for cooking the contents thoroughly. Appert patented his method, sometimes called *appertisation*.in his honor.

Appert's method was so simple and workable that it quickly became widespread. In 1810, British inventor and merchant Peter Durand, also of French origin, patented his own method, but this time in a tin can, so creating the modern-day process of canning foods. In 1812, Englishmen Bryan Donkin and John Hall purchased both patents and began producing preserves. Just a decade later, Appert's method of canning had made its way to America. Tin can production was not common until the beginning of the 20th century, partly because a hammer and chisel were needed to open cans until the invention of a can opener by an inventor named Yates in 1855.

Appert's preservation by boiling involved heating the food to an unnecessarily high temperature, and for an unnecessarily long time, which could destroy some of the flavor of the preserved food.

A less aggressive method was developed by the French chemist Louis Pasteur during an 1864 summer holiday in Arbois. To remedy the frequent acidity of the local wines, he found out experimentally that it is sufficient to heat a young wine to only about 50–60 °C (122–140 °F) for a brief time to kill the microbes, and that the wine could subsequently be aged without sacrificing the final quality. In honour of Pasteur, the process became known as "pasteurization" Pasteurization was originally used as a way of preventing wine and beer from souring, and it would be many years before milk was pasteurized. In the United States in the 1870s, it was common for milk to contain substances intended to mask spoilage before milk was regulated.

Milk

400 lbs of milk in cheese vat

Milk is an excellent medium for microbial growth, and when stored at ambient temperature bacteria and other pathogens soon proliferate.

The US Centers for Disease Control (CDC) says improperly handled raw milk is responsible for nearly three times more hospitalizations than any other food-borne disease source, making it one of the world's most dangerous food products. Diseases prevented by pasteurization can include tuberculosis, brucellosis, diphtheria, scarlet fever, and Q-fever; it also kills the harmful bacteria *Salmonella*, *Listeria*, *Yersinia*, *Campylobacter*, *Staphylococcus aureus*, and *Escherichia coli* O157:H7, among others.

Pasteurization is the reason for milk's extended shelf life. High-temperature, short-time (HTST) pasteurized milk typically has a refrigerated shelf life of two to three weeks, whereas ultra-pasteurized milk can last much longer, sometimes two to three months. When ultra-heat treatment (UHT) is combined with sterile handling and container technology (such as aseptic packaging), it can even be stored unrefrigerated for up to 9 months.

History

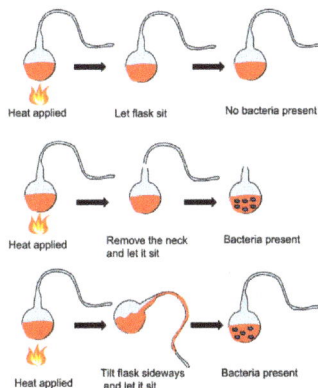

Louis Pasteur's pasteurization experiment illustrates the fact that the spoilage of liquid was caused by particles in the air rather than the air itself. These experiments were important pieces of evidence supporting the idea of Germ Theory of Disease.

Before the widespread urban growth caused by industrialization, people kept dairy cows even in urban areas and the short time period between production and consumption minimized the disease risk of drinking raw milk. As urban densities increased and supply chains lengthened to the distance from country to city, raw milk (often days old) became recognised as a source of disease. For example, between 1912 and 1937 some 65,000 people died of tuberculosis contracted from consuming milk in England and Wales alone. In the early 1900s, in Arizona, Jane H. Rider "publicized the link between infant mortality and contaminated milk, and finally convinced the dairy industry to pasteurize milk."

Developed countries adopted milk pasteurization to prevent such disease and loss of life, and as a result milk is now widely considered one of the safest foods. A traditional form of pasteurization by scalding and straining of cream to increase the keeping qualities of butter was practiced in England before 1773 and was introduced to Boston in the US by 1773, although it was not widely practiced in the United States for the next 20 years. It was still being referred to as a "new" process in American newspapers as late as 1802. Pasteurization of milk was suggested by Franz von Soxhlet in 1886. In the early 20th century, Milton Joseph Rosenau, established the standards (i.e. low temperature, slow heating at 60 °C (140 °F) for 20 minutes) for the pasteurization of milk, while at the United States Marine Hospital Service, notably in his publication of The Milk Question (1912). States in the U.S.A. began enacting mandatory dairy pasteurization laws with the first in 1947, and in 1973 the U.S. Federal Government required pasteurization of milk used in any interstate commerce.

Process

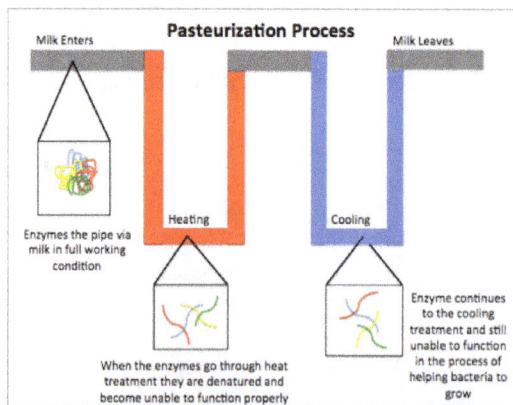

General overview of the pasteurization process. The milk starts at the left and enters the piping with functioning enzymes that, when heat treated, become denatured and stop the enzymes from functioning. This helps to stop pathogen growth by stopping the functionality of the cell. The cooling process helps stop the milk from undergoing the Maillard reaction and caramelization. The pasteurization process also has the ability to heat the cells to the point that they burst from pressure build up.

Older pasteurization methods used temperatures below boiling, since at very high temperatures, micelles of the milk protein casein irreversibly aggregate, or *curdle*. Newer methods use higher temperature, but shorten the time. Among the pasteurization methods listed below, the two main types of pasteurization used today are high-temperature, short-time (HTST, also known as "flash") and extended shelf life (ESL):

- HTST milk is forced between metal plates or through pipes heated on the outside by hot water, and the milk is heated to 72 °C (161 °F) for 15 seconds. Milk simply labeled "pasteurized" is usually treated with the HTST method.

- UHT, also known as ultra-heat-treating, processing holds the milk at a temperature of 140 °C (284 °F) for four seconds. During UHT processing milk is sterilized and not pasteurized. This process lets consumers store milk or juice for several months without refrigeration. The process is achieved by spraying the milk or juice through a nozzle into a chamber filled with high-temperature steam under pressure. After the temperature reaches 140 °C the fluid is cooled instantly in a vacuum chamber, and packed in a pre-sterilized airtight container. Milk labeled "ultra-pasteurized" or simply "UHT" has been treated with the UHT method.

- ESL milk has a microbial filtration step and lower temperatures than UHT milk. Since 2007, it is no longer a legal requirement in European countries (for example in Germany) to declare ESL milk as ultra-heated; consequently, it is now often labeled as "fresh milk" and just advertised as having an "extended shelf life," making it increasingly difficult to distinguish ESL milk from traditionally pasteurized fresh milk.

- A less conventional, but US FDA-legal, alternative (typically for home pasteurization) is to heat milk at 63 °C (145 °F) for 30 minutes.

Pasteurization methods are usually standardized and controlled by national food safety agencies (such as the USDA in the United States and the Food Standards Agency in the United Kingdom). These agencies require that milk be HTST pasteurized to qualify for the pasteurized label. Dairy product standards differ, depending on fat content and intended usage. For example, pasteurization standards for cream differ from standards for fluid milk, and standards for pasteurizing cheese are designed to preserve the enzyme phosphatase, which aids cutting. In Canada, all milk produced at a processor and intended for consumption must be pasteurized, which legally requires that it be heated to at least 72 °C for at least 16 seconds, then cooling it to 4 °C to ensure any harmful bacteria are destroyed. The UK Dairy Products Hygiene Regulations 1995 requires that milk be heat treated for 15 seconds at 71.7 °C or other effective time/temperature combination.

Some older references point to one or multiple cycles of heating and cooling (to ambient temperature or below) as either a definition of pasteurisation or a general method thereof.

A process similar to pasteurization is thermization, which uses lower temperatures to kill bacteria in milk. It allows a milk product, such as cheese, to retain more of the original taste, but thermized foods are not considered pasteurized by food regulators.

Microwave Volumetric Heating

Microwave volumetric heating (MVH) is the newest available pasteurization technology. It uses microwaves to heat liquids, suspensions, or semi-solids in a continuous flow. Because MVH delivers energy evenly and deeply into the whole body of a flowing product, it allows for gentler and shorter heating, so that almost all heat-sensitive substances in the milk are preserved.

Efficiency

The HTST pasteurization standard was designed to achieve a five-log reduction, killing 99.999% of the number of viable micro-organisms in milk. This is considered adequate for destroying almost all yeasts, molds, and common spoilage bacteria and also to ensure adequate destruction of common pathogenic, heat-resistant organisms (including *Mycobacterium tuberculosis*, which causes tuberculosis, but not *Coxiella burnetii*, which causes Q fever). As a precaution, modern equipment tests and identifies bacteria in milk being processed. HTST pasteurization processes must be designed so the milk is heated evenly, and no part of the milk is subject to a shorter time or a lower temperature.

Even pasteurization without quality control can be effective, though this is generally not permitted for human consumption; a study of farms feeding calves on pasteurized waste milk using a mixture of pasteurization technologies (none of which were routinely monitored for performance) found the resulting pasteurized milk to meet safety requirements at least 92% of the time.

An effect of the heating of pasteurization is that some vitamin, mineral, and beneficial (or probiotic) bacteria is lost. Soluble calcium and phosphorus levels decrease by 5%, thiamine (vitamin B_1) and vitamin B_{12} (cobalamin) levels by 10%, and vitamin C levels by 20%. These losses are not significant nutritionally.

Verification

Direct microbiological techniques are the ultimate measurement of pathogen contamination, but these are costly and time-consuming (24–48 hours), which means that products are able to spoil by the time pasteurization is verified.

As a result of the unsuitability of microbiological techniques, milk pasteurization efficacy is typically monitored by checking for the presence of alkaline phosphatase, which is denatured by pasteurization. B. tuberculosis, the bacterium that requires the highest temperature to be killed of all milk pathogens is killed at ranges of temperature and time similar to those that denature alkaline phosphatase. For this reason, presence of alkaline phosphatase is an ideal indicator of pasteurization efficacy.

Phosphatase denaturing was originally monitored using a phenol-phosphate substrate. When hydrolysed by the enzyme these compounds liberate phenols, which were then reacted with dibromoquinonechlorimide to give a colour change, which itself was measured by checking absorption at 610 nm (spectrophotometry). Some of the phenols used were inherently coloured (phenolpthalein, nitrophenol) and were simply assayed unreacted. Spectrophotometric analysis is satisfactory but is of relatively low accuracy because many natural products are coloured. For this reason, modern systems (since 1990) use fluorometry which is able to detect much lower levels of raw milk contamination.

Unpasteurized Milk

According to the United States Centers for Disease Control between 1998 and 2011 79% of the dairy related outbreaks were due to raw milk or cheese products. They report 148 outbreaks, 2,384 illnesses (284 requiring hospitalizations) as well as 2 deaths due to raw milk or cheese products during the same time period.

Consumer Acceptance

Although pasteurization has been practiced for a long time, some consumers contend that they should have the right to buy and sell unpasteurized milk if they want to.

Some consumers also point out that government-enforced pasteurization law has been used as a tool for large business to shut out competition from smaller producers.

Fouling

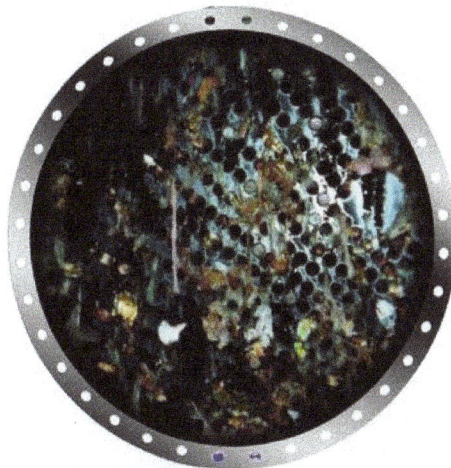

Heat exchanger in a steam power plant, fouled by macro fouling

Condenser tube with residues of biofouling (cut open)

Fouling is the accumulation of unwanted material on solid surfaces to the detriment of function. The fouling materials can consist of either living organisms (biofouling) or a non-living substance (inorganic or organic). Fouling is usually distinguished from other surface-growth phenomena, in that it occurs on a surface of a component, system or plant performing a defined and useful function, and that the fouling process impedes or interferes with this function.

Other terms used in the literature to describe fouling include: deposit formation, encrustation, crudding, deposition, scaling, scale formation, slagging, and sludge formation. The last six terms have a more narrow meaning than fouling within the scope of the fouling science and technology, and they also have meanings outside of this scope; therefore, they should be used with caution.

Fouling phenomena are common and diverse, ranging from fouling of ship hulls, natural surfaces in the marine environment (marine fouling), fouling of heat-transfer components through ingredients contained in the cooling water or gases, and even the development of plaque or calculus on teeth, or deposits on solar panels on Mars, among other examples.

This article is primarily devoted to the fouling of industrial heat exchangers, although the same theory is generally applicable to other varieties of fouling. In the cooling technology and other technical fields, a distinction is made between macro fouling and micro fouling. Of the two, micro fouling is the one which is usually more difficult to prevent and therefore more important.

Components Subject to Fouling

Following are examples of components that may be subject to fouling and the corresponding effects of fouling:

- Heat exchanger surfaces – reduces thermal efficiency, decreases heat flux, increases temperature on the hot side, decreases temperature on the cold side, induces under-deposit corrosion, increases use of cooling water;

- Piping, flow channels – reduces flow, increases pressure drop, increases upstream pressure, increases energy expenditure, may cause flow oscillations, slugging in two-phase flow, cavitation; may increase flow velocity elsewhere, may induce vibrations, may cause flow blockage;

- Ship hulls – creates additional drag, increases fuel usage, reduces maximum speed;

- Turbines – reduces efficiency, increases probability of failure;

- Solar panels – decreases the electrical power generated;

- Reverse osmosis membranes – increases pressure drop, increases energy expenditure, reduces flux, membrane failure (in severe cases);

- Electrical heating elements – increases temperature of the element, increases corrosion, reduces lifespan;

- Nuclear fuel in pressurized water reactors – axial offset anomaly, may need to de-rate the power plant;

- Injection/spray nozzles (e.g., a nozzle spraying a fuel into a furnace) – incorrect amount injected, malformed jet, component inefficiency, component failure;

- Venturi tubes, orifice plates – inaccurate or incorrect measurement of flow rate;

- Pitot tubes in airplanes – inaccurate or incorrect indication of airplane speed;

- Spark plug electrodes in cars – engine misfiring;

- Production zone of petroleum reservoirs and oil wells – decreased petroleum production with time; plugging; in some cases complete stoppage of flow in a matter of days;

- Teeth – promotes tooth or gum disease, decreases aesthetics;

- Living organisms – deposition of excess minerals (e.g., calcium, iron, copper) in tissues is (sometimes controversially) linked to aging/senescence.

Macro Fouling

Macro fouling is caused by coarse matter of either biological or inorganic origin, for example industrially produced refuse. Such matter enters into the cooling water circuit through the cooling water pumps from sources like the open sea, rivers or lakes. In closed circuits, like cooling towers, the ingress of macro fouling into the cooling tower basin is possible through open canals or by the wind. Sometimes, parts of the cooling tower internals detach themselves and are carried into the cooling water circuit. Such substances can foul the surfaces of heat exchangers and may cause deterioration of the relevant heat transfer coefficient. They may also create flow blockages, redistribute the flow inside the components, or cause fretting damage.

Examples

- Manmade refuse;

- Detached internal parts of components;
- Tools and other "foreign objects" accidentally left after maintenance;
- Algae;
- Mussels;
- Leaves, parts of plants up to entire trunks.

Micro Fouling

As to micro fouling, distinctions are made between:

- Scaling or precipitation fouling, as crystallization of solid salts, oxides and hydroxides from water solutions, for example, calcium carbonate or calcium sulfate;
- Particulate fouling, i.e., accumulation of particles, typically colloidal particles, on a surface;
- Corrosion fouling, i.e., in-situ growth of corrosion deposits, for example, magnetite on carbon steel surfaces;
- Chemical reaction fouling, for example, decomposition or polymerization of organic matter on heating surfaces;
- Solidification fouling - when components of the flowing fluid with a high-melting point freeze onto a subcooled surface;
- Biofouling, like settlements of bacteria and algae;
- Composite fouling, whereby fouling involves more than one foulant or fouling mechanism.

Precipitation Fouling

Limescale buildup inside a pipe both reduces liquid flow through the pipe, as well as reduces thermal conduction from the liquid to the outer pipe shell. Both effects will reduce the pipe's overall thermal efficiency when used as a heat exchanger.

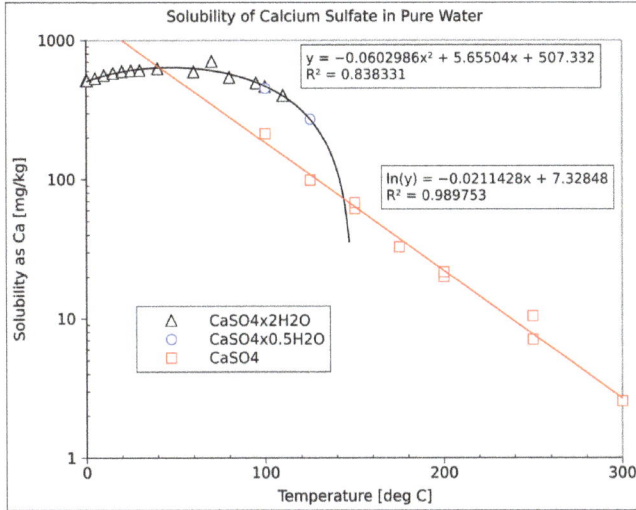

Temperature dependence of the solubility of calcium sulfate (3 phases) in pure water. The water is pressurized so that it can be maintained in the liquid state at the elevated temperatures.

Scaling or precipitation fouling involves crystallization of solid salts, oxides and hydroxides from solutions. These are most often water solutions, but non-aqueous precipitation fouling is also known. Precipitation fouling is a very common problem in boilers and heat exchangers operating with hard water and often results in limescale.

Through changes in temperature, or solvent evaporation or degasification, the concentration of salts may exceed the saturation, leading to a precipitation of solids (usually crystals).

As an example, the equilibrium between the readily soluble calcium bicarbonate - always prevailing in natural water - and the poorly soluble calcium carbonate, the following chemical equation may be written:

$$Ca(HCO3)2_{(aqueous)} -> CaCO3\downarrow + CO2\uparrow + H2O$$

The calcium carbonate that forms through this reaction precipitates. Due to the temperature dependence of the reaction, and increasing volatility of CO_2 with increasing temperature, the scaling is higher at the hotter outlet of the heat exchanger than at the cooler inlet.

In general, the dependence of the salt solubility on temperature or presence of evaporation will often be the driving force for precipitation fouling. The important distinction is between salts with "normal" or "retrograde" dependence of solubility on temperature. The salts with the "normal" solubility increase their solubility with increasing temperature and thus will foul the cooling surfaces. The salts with "inverse" or "retrograde" solubility will foul the heating surfaces. An example of the temperature dependence of solubility is shown in the figure. Calcium sulfate is a common precipitation foulant of heating surfaces due to its retrograde solubility.

Precipitation fouling can also occur in the absence of heating or vaporization. For example, calcium sulfate decreases it solubility with decreasing pressure. This can lead to precipitation fouling of reservoirs and wells in oil fields, decreasing their productivity with time. Fouling of membranes in reverse osmosis systems can occur due to differential solubility of barium sulfate in solutions of different ionic strength. Similarly, precipitation fouling can occur because of solubility changes induced by other factors, e.g., liquid flashing, liquid degassing, redox potential changes, or mixing of incompatible fluid streams.

The following lists some of the industrially common phases of precipitation fouling deposits observed in practice to form from aqueous solutions:

- Calcium carbonate (calcite, aragonite usually at t > ~50 °C, or rarely vaterite);

- Calcium sulfate (anhydrite, hemihydrate, gypsum);

- Calcium oxalate (e.g., beerstone);

- Barium sulfate (barite);

- Magnesium hydroxide (brucite); magnesium oxide (periclase);

- Silicates (serpentine, acmite, gyrolite, gehlenite, amorphous silica, quartz, cristobalite, pectolite, xonotlite);

- Aluminium oxide hydroxides (boehmite, gibbsite, diaspore, corundum);

- Aluminosilicates (analcite, cancrinite, noselite);

- Copper (metallic copper, cuprite, tenorite);

- Phosphates (hydroxyapatite);

- Magnetite or nickel ferrite ($NiFe_2O_4$) from extremely pure, low-iron water.

The deposition rate by precipitation is often described by the following equations:

Transport: $\dfrac{dm}{dt} = k_t(C_b - C_i)$

Surface crystallisation: $\dfrac{dm}{dt} = k_r(C_i - C_e)^{n1}$

Overall: $\dfrac{dm}{dt} = k_d(C_b - C_e)^{n2}$

where:

m - mass of the material (per unit surface area), kg/m²

t - time, s

C_b - concentration of the substance in the bulk of the fluid, kg/m³

C_i - concentration of the substance at the interface, kg/m³

C_e - equilibrium concentration of the substance at the conditions of the interface, kg/m³

n1, n2 - order of reaction for the crystallisation reaction and the overall deposition process, respectively, dimensionless

k_t, k_r, k_d - kinetic rate constants for the transport, the surface reaction, and the overall deposition reaction, respectively; with the dimension of m/s (when n1 and n2 = 1)

Particulate Fouling

Fouling by particles suspended in water ("crud") or in gas progresses by a mechanism different than precipitation fouling. This process is usually most important for colloidal particles, i.e., particles smaller than about 1 µm in at least one dimension (but which are much larger than atomic dimensions). Particles are transported to the surface by a number of mechanisms and there they can attach themselves, e.g., by flocculation or coagulation. Note that the attachment of colloidal particles typically involves electrical forces and thus the particle behaviour defies the experience from the macroscopic world. The probability of attachment is sometimes referred to as "sticking probability", P:

$$k_d = Pk_t$$

where k_d and k_t are the kinetic rate constants for deposition and transport, respectively. The value of P for colloidal particles is a function of both the surface chemistry, geometry, and the local thermohydraulic conditions.

An alternative to using the sticking probability is to use a kinetic attachment rate constant, assuming the first order reaction:

$$\frac{dm}{dt} = k_a C_i$$

and then the transport and attachment kinetic coefficients are combined as two processes occurring in series:

$$k_d = \left(\frac{1}{k_a} + \frac{1}{k_t} \right)^{-1}$$

$$\frac{dm}{dt} = k_d C_b$$

where:

- dm/dt is the rate of the deposition by particles, kg m⁻² s⁻¹,

- k_a, k_t and k_d are the kinetic rate constants for deposition, m/s,

- C_i and C_b are the concentration of the particle foulant at the interface and in the bulk fluid, respectively; kg m³.

Being essentially a surface chemistry phenomenon, this fouling mechanism can be very sensitive to factors that affect colloidal stability, e.g., zeta potential. A maximum fouling rate is usually observed when the fouling particles and the substrate exhibit opposite electrical charge, or near the point of zero charge of either of them.

Particles larger than those of colloidal dimensions may also foul e.g., by sedimentation ("sedimentation fouling") or straining in small-size openings.

With time, the resulting surface deposit may harden through processes collectively known as "deposit consolidation" or, colloquially, "aging".

The common particulate fouling deposits formed from aqueous suspensions include:

- iron oxides and iron oxyhydroxides (magnetite, hematite, lepidocrocite, maghemite, goethite);

- Sedimentation fouling by silt and other relatively coarse suspended matter.

Fouling by particles from gas aerosols is also of industrial significance. The particles can be either solid or liquid. The common examples can be fouling by flue gases, or fouling of air-cooled components by dust in air. The mechanisms are discussed in article on aerosol deposition.

Corrosion Fouling

Corrosion deposits are created in-situ by the corrosion of the substrate. They are distinguished from fouling deposits, which form from material originating ex-situ. Corrosion deposits should not be confused with fouling deposits formed by ex-situ generated corrosion products. Corrosion deposits will normally have composition related to the composition of the substrate. Also, the geometry of the metal-oxide and oxide-fluid interfaces may allow practical distinction between the corrosion and fouling deposits. An example of corrosion fouling can be formation of an iron oxide or oxyhydroxide deposit from corrosion of the carbon steel underneath. Corrosion fouling should not be confused with fouling corrosion, i.e., any of the types of corrosion that may be induced by fouling.

Chemical Reaction Fouling

Chemical reactions may occur on contact of the chemical species in the process fluid with heat transfer surfaces. In such cases, the metallic surface sometimes acts as a catalyst. For example, corrosion and polymerization occurs in cooling water for the chemical industry which has a minor content of hydrocarbons. Systems in petroleum processing are prone to polymerization of olefins or deposition of heavy fractions (asphaltenes, waxes, etc.). High tube wall temperatures may lead to carbonizing of organic matter. Food industry, for example milk processing, also experiences fouling problems by chemical reactions.

Fouling through an ionic reaction with an evolution of an inorganic solid is commonly classified as precipitation fouling (not chemical reaction fouling).

Solidification Fouling

Solidification fouling occurs when a component of the flowing fluid "freezes" onto a surface forming a solid fouling deposit. Examples may include solidification of wax (with a high melting point) from a hydrocarbon solution, or of molten ash (carried in a furnace exhaust gas) onto a heat exchanger surface. The surface needs to have a temperature below a certain threshold; therefore, it is said to be subcooled in respect to the solidification point of the foulant.

Biofouling

A fragment of a canal lock in Northern France, covered with zebra mussels

Biofouling or biological fouling is the undesirable accumulation of micro-organisms, algae and diatoms, plants, and animals on surfaces, for example ships' hulls, or piping and reservoirs with untreated water. This can be accompanied by microbiologically influenced corrosion (MIC).

Bacteria can form biofilms or slimes. Thus the organisms can aggregate on surfaces using colloidal hydrogels of water and extracellular polymeric substances (EPS) (polysaccharides, lipids, nucleic acids, etc.). The biofilm structure is usually complex.

Bacterial fouling can occur under either aerobic (with oxygen dissolved in water) or anaerobic (no oxygen) conditions. In practice, aerobic bacteria prefer open systems, when both oxygen and nutrients are constantly delivered, often in warm and sunlit environments. Anaerobic fouling more often occurs in closed systems when sufficient nutrients are present. Examples may include sulfate-reducing bacteria (or sulfur-reducing bacteria), which produce sulfide and often cause corrosion of ferrous metals (and other alloys). Sulfide-oxidizing bacteria (e.g., Acidithiobacillus), on the other hand, can produce sulfuric acid, and can be involved in corrosion of concrete.

Zebra mussels serve as an example of larger animals that have caused widespread fouling in North America.

Composite Fouling

Composite fouling is common. This type of fouling involves more than one foulant or more than one fouling mechanism working simultaneously. The multiple foulants or mechanisms may interact with each other resulting in a synergistic fouling which is not a simple arithmetic sum of the individual components.

Fouling on Mars

NASA Mars Exploration Rovers (Spirit and Opportunity) experienced (presumably) abiotic fouling of solar panels by dust particles from the Martian atmosphere. Some of the deposits subsequently spontaneously cleaned off. This illustrates the universal nature of the fouling phenomena.

Quantification of Fouling

The most straightforward way to quantify fairly uniform fouling is by stating the average deposit surface loading, i.e., kg of deposit per m^2 of surface area. The fouling rate will then be expressed in kg/m^{2}s, and it is obtained by dividing the deposit surface loading by the effective operating time. The normalized fouling rate (also in kg/m^{2}s) will additionally account for the concentration of the foulant in the process fluid (kg/kg) during preceding operations, and is useful for comparison of fouling rates between different systems. It is obtained by dividing the fouling rate by the foulant concentration. The fouling rate constant (m/s) can be obtained by dividing the normalized fouling rate by the mass density of the process fluid (kg/m^3).

Deposit thickness (μm) and porosity (%) are also often used for description of fouling amount. The relative reduction of diameter of piping or increase of the surface roughness can be of particular interest when the impact of fouling on pressure drop is of interest.

In heat transfer equipment, where the primary concern is often the effect of fouling on heat transfer, fouling can be quantified by the increase of the resistance to the flow of heat (m^{2}K/W) due to fouling (termed "fouling resistance"), or by development of heat transfer coefficient (W/m^{2}K) with time.

If under-deposit or crevice corrosion is of primary concern, it is important to note non-uniformity of deposit thickness (e.g., deposit waviness), localized fouling, packing of confined regions with deposits, creation of occlusions, "crevices", "deposit tubercles", or sludge piles. Such deposit structures can create environment for underdeposit corrosion of the substrate material, e.g., intergranular attack, pitting, stress corrosion cracking, or localized wastage. Porosity and permeability of the deposits will likely influence the probability of underdeposit corrosion. Deposit composition can also be important - even minor components of the deposits can sometimes cause severe corrosion of the underlying metal (e.g., vanadium in deposits of fired boilers causing hot corrosion).

There is no general rule on how much deposit can be tolerated, it depends on the system. In many cases, a deposit even a few micrometers thick can be troublesome. A deposit in a millimeter-range thickness will be of concern in almost any application.

Progress of Fouling with Time

Deposit on a surface does not always develop steadily with time. The following fouling scenarios can be distinguished, depending on the nature of the system and the local thermohydraulic conditions at the surface:

- Induction period. Sometimes, a near-nil fouling rate is observed when the surface is new or very clean. This is often observed in biofouling and precipitation fouling. After the "induction period", the fouling rate increases.

- "Negative" fouling. This can occur when fouling rate is quantified by monitoring heat transfer. Relatively small amounts of deposit can improve heat transfer, relative to clean surface, and give an appearance of "negative" fouling rate and negative total fouling amount. Negative fouling is often observed under nucleate-boiling heat-transfer conditions (deposit improves bubble nucleation) or forced-convection (if the deposit increases the surface roughness and the surface is no longer "hydraulically smooth"). After the initial period of "surface roughness control", the fouling rate usually becomes strongly positive.

- Linear fouling. The fouling rate can be steady with time. This is a common case.

- Falling fouling. Under this scenario, the fouling rate decreases with time, but never drops to zero. The deposit thickness does not achieve a constant value. The progress of fouling can be often described by two numbers: the initial fouling rate (a tangent to the fouling curve at zero deposit loading or zero time) and the fouling rate after a long period of time (an oblique asymptote to the fouling curve).

- Asymptotic fouling. Here, the fouling rate decreases with time, until it finally reaches zero. At this point, the deposit thickness remains constant with time (a horizontal asymptote). This is often the case for relatively soft or poorly adherent deposits in areas of fast flow. The asymptote is usually interpreted as the deposit loading at which the deposition rate equals the deposit removal rate.

- Accelerating fouling. Under this scenario, the fouling rate increases with time; the rate of deposit buildup accelerates with time (perhaps until it becomes transport limited). Mechanistically, this scenario can develop when fouling increases the surface roughness, or when the deposit surface exhibits higher chemical propensity to fouling than the pure underlying metal.

- Seesaw fouling. Here, fouling loading generally increases with time (often assuming a generally linear or falling rate), but, when looked at in more detail, the fouling progress is periodically interrupted and takes the form of sawtooth curve. The periodic sharp variations in the apparent fouling amount often correspond to the moments of system shutdowns, startups or other transients in operation. The periodic variations are often interpreted as periodic removal of some of the deposit (perhaps deposit re-suspension due to pressure pulses, spalling due thermal stresses, or exfoliation due to redox transients). Steam blanketing has been postulated to occur between the partially spalled deposits and the heat transfer surface. However, other reasons are possible, e.g., trapping of air inside the surface deposits during shutdowns, or inaccuracy of temperature measurements during transients ("temperature streaming").

Fouling Modelling

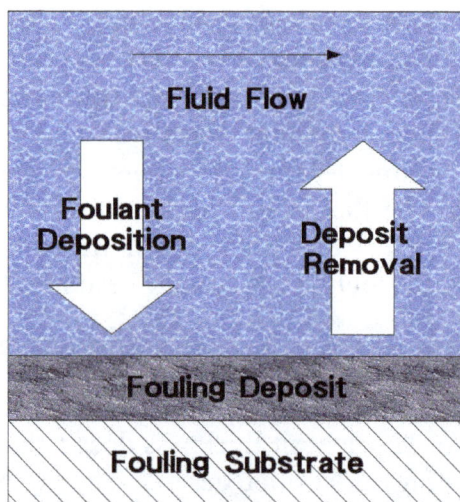

Schematics of the fouling process consisting of simultaneous foulant deposition and deposit removal.

Fouling of a system can be modelled as consisting of several steps:

- Generation or ingress of the species that causes fouling ("foulant sourcing");

- Foulant transport with the stream of the process fluid (most often by advection);

- Foulant transport from the bulk of the process fluid to the fouling surface. This transport is often by molecular or turbulent-eddy diffusion, but may also occur by inertial coasting/impaction, particle interception by the surface (for particles with finite sizes), electrophoresis, thermophoresis, diffusiophoresis, Stefan flow (in condensation and evaporation), sedimentation, Magnus force (acting on rotating particles), thermoelectric effect, and other mechanisms.

- Induction period, i.e., a near-nil fouling rate at the initial period of fouling (observed only for some fouling mechanisms);

- Foulant crystallisation on the surface (or attachment of the colloidal particle, or chemical reaction, or bacterial growth);

- Sometimes fouling autoretardation, i.e., reduction (or potentially enhancement) of crystallisation/attachment rate due to changes in the surface conditions caused by the fouling deposit;

- Deposit dissolution (or re-entrainment of loosely attached particles);

- Deposit consolidation on the surface (e.g., through Ostwald ripening or differential solubility in temperature gradient) or cementation, which account for deposit losing its porosity and becoming more tenacious with time;

- Deposit spalling, erosion wear, or exfoiliation.

Deposition consists of transport to the surface and subsequent attachment. Deposit removal is either through deposit dissolution, particle re-entrainment, or deposit spalling, erosive wear, or exfoliation. Fouling results from foulant generation, foulant deposition, deposit removal, and deposit consolidation.

For the modern model of fouling involving deposition with simultaneous deposit re-entrainment and consolidation, the fouling process can be represented by the following scheme:

[rate of deposit accumulation] = [rate of deposition] - [rate of re-entrainment of unconsolidated deposit]

[rate of accumulation of unconsolidated deposit] = [rate of deposition] - [rate of re-entrainment of unconsolidated deposit] - [rate of consolidation of unconsolidated deposit]

Following the above scheme, the basic fouling equations can be written as follows (for steady-state conditions with flow, when concentration remains constant with time):

$$\begin{cases} dm/dt = k_d C_m \rho - \lambda_r m_r(t) \\ dm_r/dt = k_d C_m \rho - \lambda_r m_r(t) - \lambda_c \cdot m_r(t) \end{cases}$$

where:

- m is the mass loading of the deposit (consolidated and unconsolidated) on the surface (kg/m²);

- t is time (s);

- k_d is the deposition rate constant (m/s);

- ρ is the fluid density (kg/m³);

- C_m - mass fraction of foulant in the fluid (kg/kg);

- λ_r is the re-entrainment rate constant (1/s);

- m_r is the mass loading of the removable (i.e., unconsolidated) fraction of the surface deposit (kg/m²); and

- λ_c is the consolidation rate constant (1/s).

This system of equations can be integrated (taking that m = 0 and m_r = 0 at t = 0) to the form:

$$m(t) = \frac{k_d C_m \rho}{\lambda}\left(t\lambda_c + \frac{\lambda_r}{\lambda}\left(1 - e^{-\lambda t}\right)\right)$$

where $\lambda = \lambda_r + \lambda_c$.

This model reproduces either linear, falling, or asymptotic fouling, depending on the relative values of k, λ_r, and λ_c. The underlying physical picture for this model is that of a two-layer deposit consisting of consolidated inner layer and loose unconsolidated outer layer. Such a bi-layer deposit is often observed in practice. The above model simplifies readily to the older model of simultaneous deposition and re-entrainment (which neglects consolidation) when λ_c=0. In the absence of consolidation, the asymptotic fouling is always anticipated by this older model and the fouling progress can be described as:

$$m(t) = m^*\left(1 - e^{-\lambda_r t}\right)$$

where m* is the maximum (asymptotic) mass loading of the deposit on the surface (kg/m²).

Economic and Environmental Importance of Fouling

Fouling is ubiquitous and generates tremendous operational losses, not unlike corrosion. For example, one estimate puts the losses due to fouling of heat exchangers in industrialized nations to be about 0.25% of their GDP. Another analysis estimated (for 2006) the economical loss due to boiler and turbine fouling in China utilities at 4.68 billion dollars, which is about 0.169% the country GDP .

The losses initially result from impaired heat transfer, corrosion damage (in particular under-deposit and crevice corrosion), increased pressure drop, flow blockages, flow redistribution inside components, flow instabilities, induced vibrations (possibly leading to other problems, e.g., fatigue), fretting, premature failure of electrical heating elements, and a large number of other often unanticipated problems. In addition, the ecological costs should be (but typically are not) considered. The ecological costs arise from the use of biocides for the avoidance of biofouling, from the increased fuel input to compensate for the reduced output caused by fouling, and an increased use of cooling water in once-through cooling systems.

For example, "normal" fouling at a conventionally fired 500 MW (net electrical power) power station unit accounts for output losses of the steam turbine of 5 MW and more. In a 1,300 MW nuclear power station, typical losses could be 20 MW and up (up to 100% if the station shuts down due to fouling-induced component degradation). In seawater desalination plants, fouling may reduce the gained output ratio by two-digit percentages (the gained output ratio is an equivalent that puts the mass of generated distillate in relation to the steam used in the process). The extra electrical consumption in compressor-operated coolers is also easily in the two-digit area. In addition to the operational costs, also the capital cost increases because the heat exchangers have to be designed in larger sizes to compensate for the heat-transfer loss due to fouling. To the output losses listed above, one needs to add the cost of down-time required to inspect, clean, and repair the components (millions of dollars per day of shutdown in lost revenue in a typical power plant), and the cost of actually doing this maintenance. Finally, fouling is often a root cause of serious degradation problems that may limit the life of components or entire plants.

Fouling Control

The most fundamental and usually preferred method of controlling fouling is to prevent the ingress of the fouling species into the cooling water circuit. In steam power stations and other major industrial installations of water technology, macro fouling is avoided by way of pre-filtration and cooling water debris filters. Some plants employ foreign-object exclusion program (to eliminate the possibility of salient introduction of unwanted materials, e.g., forgetting tools during maintenance). Acoustic monitoring is sometimes employed to monitor for fretting by detached parts. In the case of micro fouling, water purification is achieved with extensive methods of water treatment, microfiltration, membrane technology (reverse osmosis, electrodeionization) or ion-exchange resins. The generation of the corrosion products in the water piping systems is often minimized by controlling the pH of the process fluid (typically alkanization with ammonia, morpholine, ethanolamine or sodium phosphate), control of oxygen dissolved in water (for example, by addition of hydrazine), or addition of corrosion inhibitors.

For water systems at relatively low temperatures, the applied biocides may be classified as follows: inorganic chlorine and bromide compounds, chlorine and bromide cleavers, ozone and oxygen cleavers, unoxidizable biocides. One of the most important unoxidizable biocides is a mixture of chloromethyl-isothiazolinone and methyl-isothiazolinone. Also applied are dibrom nitrilopropionamide and quaternary ammonium compounds. For underwater ship hulls bottom paints are applied.

Chemical fouling inhibitors can reduce fouling in many systems, mainly by interfering with the crystallization, attachment, or consolidation steps of the fouling process. Examples for water systems are: chelating agents (for example, EDTA), long-chain aliphatic amines or polyamines (for example, octadecylamine, helamin, and other "film-forming" amines), organic phosphonic acids (for example, etidronic acid), or polyelectrolytes (for example, polyacrylic acid, polymethacrylic acid, usually with a molecular weight lower than 10000). For fired boilers, aluminum or magnesium additives can lower the melting point of ash and promote creation of deposits which are easier to remove.

Magnetic water treatment has been a subject of controversy as to its effectiveness for fouling control since the 1950s. The prevailing opinion is that it simply "does not work". Nevertheless, some studies suggest that it may be effective under some conditions to reduce buildup of calcium carbonate deposits.

On the component design level, fouling can often (but not always) be minimized by maintaining a relatively high (for example, 2 m/s) and uniform fluid velocity throughout the component. Stagnant regions need to be eliminated. Components are normally overdesigned to accommodate the fouling anticipated between cleanings. However, a significant overdesign can be a design error because it may lead to increased fouling due to reduced velocities. Periodic on-line pressure pulses or backflow can be effective if the capability is carefully incorporated at the design time. Blowdown capability is always incorporated into steam generators or evaporators to control the accumulation of non-volatile impurities that cause or aggravate fouling. Low-fouling surfaces (for example, very smooth, implanted with ions, or of low surface energy like Teflon) are an option for some applications. Modern components are typically required to be designed for ease of inspection of internals and periodic cleaning. On-line fouling monitoring systems are designed for some application so that blowing or cleaning can be applied before unpredictable shutdown is necessary or damage occurs.

Chemical or mechanical cleaning processes for the removal of deposits and scales are recommended when fouling reaches the point of impacting the system performance or an onset of significant fouling-induced degradation (e.g., by corrosion). These processes comprise pickling with acids and complexing agents, cleaning with high-velocity water jets ("water lancing"), recirculating ("blasting") with metal, sponge or other balls, or propelling offline mechanical "bullet-type" tube cleaners. Whereas chemical cleaning causes environmental problems through the handling, application, storage and disposal of chemicals, the mechanical cleaning by means of circulating cleaning balls or offline "bullet-type" cleaning can be an environmentally friendlier alternative. In some heat-transfer applications, mechanical mitigation with dynamic scraped surface heat exchangers is an option. Also ultrasonic or abrasive cleaning methods are available for many specific applications.

Soured Milk

Soured milk

Soured milk is a food product produced by the acidification of milk. Acidification, which gives the milk a tart taste, is achieved either through the addition of an acid, such as lemon juice or vinegar, or through bacterial fermentation. The acid causes milk to coagulate and thicken, inhibiting the growth of harmful bacteria and improving the product's shelf life. Soured milk that is produced by bacterial fermentation is more specifically called fermented milk or cultured milk.

Modern commercial soured milk differs from milk that has become sour naturally, though the latter is also commonly known as "soured milk." Traditionally, soured milk was simply fresh milk that was left to ferment and sour by keeping it in a warm place for a day, often near a stove. The milk would become sour and then ferment.

Soured milk that is produced by the addition of an acid, with or without the addition of microbial organisms, is more specifically called acidified milk. In the United States, acids used to manufacture acidified milk include acetic acid (commonly found in vinegar), adipic acid, citric acid (commonly found in lemon juice), fumaric acid, glucono-delta-lactone, hydrochloric acid, lactic acid, malic acid, phosphoric acid, succinic acid, and tartaric acid.

Soured milk is commonly made at home or is sold and consumed in Europe, especially in Eastern Europe (Belarus, Poland, Slovakia, Russia, Ukraine), all over the countries of the former Yugoslavia (Macedonia, Serbia, Montenegro, Bosnia and Herzegovina, Croatia, Slovenia), Bulgaria, Finland, Germany, and Scandinavia.

It is also made at home or sold in supermarkets and consumed in the Great Lakes region of Eastern Africa (Kenya, Uganda, Rwanda, Burundi and Tanzania). It is also a traditional food of the Bantu people of Southern Africa.

Since the 1970s, some producers have used chemical acidification in place of biological agents.

In recipes

Raw milk that has not gone sour is sometimes referred to as "sweet milk," because it contains the sugar lactose. Fermentation converts the lactose to lactic acid, which has a sour flavor. Before refrigeration, raw milk commonly became sour before it could be consumed, and various recipes incorporate such leftover milk as an ingredient. Sour milk produced by fermentation differs in flavor from that produced by acidification, because the acids commonly added in commercial manufacture, such as acetic acid and citric acid, have different flavors from lactic acid, and also because fermentation can introduce new flavors. Modern food safety standards mean that milk that has gone sour is considered safe to drink.

Buttermilk is a common modern substitute for naturally soured milk.

Raw Milk

Raw milk is milk that has not been pasteurized. While proponents have stated that there are benefits to consuming raw milk, the medical community has warned of the dangers of consuming unpasteurized milk. Availability and regulation of raw milk vary from region to region.

History of Raw Milk and Pasteurization

Cattle have been domesticated for some 10,500 years. Europeans first started consuming milk from cattle around 7,500 years ago.

Pasteurization is widely used to prevent infected milk from entering the food supply. The Pasteurization process was developed in 1864 by French scientist Louis Pasteur, who discovered that heating beer and wine was enough to kill most of the bacteria that caused spoilage, preventing these beverages from turning sour. The process achieves this by eliminating pathogenic microbes and lowering microbial numbers to prolong the quality of the beverage.

After sufficient scientific study led to the development of germ theory, pasteurization was introduced in the United States in the 1890s. This move successfully controlled the spread of highly contagious bacterial diseases including *E. coli*, bovine tuberculosis and brucellosis, (all thought to be easily transmitted to humans through the drinking of raw milk). In the early days after the scientific discovery of bacteria, there was no product testing to determine whether a farmer's milk was safe or infected, so all milk was treated as potentially contagious. After the first tests were developed, some farmers took steps to prevent their infected animals from being killed and removed from food production, sometimes even falsifying test results to make their animals appear free of infection. Recent advances in the analysis of milk-borne diseases have enabled scientists to track the DNA of the infectious bacteria to the cows on the farms that supplied the raw milk.

The recognition of many potentially deadly pathogens, such as *E. coli* 0157 H7, *Campylobacter*, *Listeria*, and *Salmonella*, and their presence in milk products has led to the continuation of pasteurization. The Department of Health and Human Services, Center for Disease Control and Prevention, and other health agencies of the United States strongly recommend that the public do not consume raw milk or raw milk products. Young children, the elderly, people with weakened immune systems, and pregnant women are particularly susceptible to infections originating in raw milk.

In the USA, re-pasteurization occurs when pasteurized milk from the US mainland is transported by sea to Hawaii, and then pasteurized again.

Raw vs. Pasteurized Debate

Those favoring the consumption of raw milk believe that raw milk and associated products are healthier and taste better. Those favoring the consumption of pasteurized milk consider the pathogen risk associated with drinking raw milk unacceptable.

Agencies such as the Centers for Disease Control and Prevention (CDC), and the Food and Drug Administration (FDA) in the United States, and other regulatory agencies around the world say that pathogens from raw milk, including potentially tuberculosis, diphtheria, typhoid, and streptococcal infections, make it unsafe to consume. Similarly, a recent review authored by the Belgian Federal Agency for the Safety of the Food Chain and experts from Belgian universities and institutions concluded that "raw milk poses a realistic health threat due to a possible contamination with human pathogens. It is therefore strongly recommended that milk should be heated before consumption. With the exception of an altered organoleptic [flavor] profile, heating (particularly ultra high temperature and similar treatments) will not substantially change the nutritional value of raw milk or other benefits associated with raw milk consumption."

Raw milk advocates, such as the Weston A. Price Foundation, say that raw milk can be produced hygienically, and that it has health benefits that are destroyed in the pasteurization process. Research shows only very slight differences in the nutritional values of pasteurized and unpasteurized milk.

Three studies have found a statistically significant inverse relationship between consumption of raw milk and asthma and allergies. However, all of these studies have been performed in children living on farms and living a farming lifestyle, rather than comparing urban children living typical urban lifestyles and with typical urban exposures on the basis of consumption or nonconsumption of raw milk. Aspects of the overall urban vs. farming environment lifestyle have been suggested as having a role in these differences, and for this reason, the overall phenomenon has been named the "farm effect." A recent scientific review concluded that "most studies alluding to a possible protective effect of raw milk consumption do not contain any objective confirmation of the raw milk's status or a direct comparison with heat-treated milk. Moreover, it seems that the observed increased resistance seems to be rather related to the exposure to a farm environment or to animals than to raw milk consumption." For example, in the largest of these studies, exposure to cows and straw as well as raw milk were associated with lower rates of asthma, and exposure to animal feed storage rooms and manure with lower rates of atopic dermatitis; "the effect on hay fever and atopic sensitization could not be completely explained by the questionnaire items themselves or their diversity."

Legal Status

Regulation of the commercial distribution of packaged raw milk varies across the world. Some countries have complete bans, but many had partial bans that do not restrict the purchase of raw milk bought directly from the farmer. Raw milk is sometimes distributed through a program, in which the consumer owns a share in the dairy animal or the herd, and therefore can be considered to be consuming milk from their own animal, which is legal. Raw milk is sometimes marketed for animal or pet consumption, or for other uses such as soap making in places where sales for human consumption are prohibited.

Africa

Although milk consumption in Africa is fairly low compared to the rest of the world, in tribes where milk consumption is popular, such as the Maasai tribe, milk is typically consumed unpasteurized.

Asia

In rural areas of Asia where milk consumption is popular, milk is typically unpasteurized. In large cities of Asia, raw milk, especially from water buffalo, is typical. In most countries of Asia, laws prohibiting raw milk are nonexistent or rarely enforced.

Europe

The European Union requires that raw milk and products made with raw milk must be labeled to indicate this. Under EU hygiene rules, member states can prohibit or restrict the placing on the market of raw milk intended for human consumption. Also, European countries are free to add certain requirements. Usually special sanitary regulations and frequent quality tests (at least once per month) are mandatory.

French Roquefort, a famous blue cheese, which is required by European law to be made from raw sheep's milk.

France

Raw-milk cheeses make up about 18 percent of France's total cheese production, and are considered far superior to pasteurised cheeses. Many French cuisine traditionalists consider pasteurized cheeses almost a sacrilege. Many traditional French cheeses have solely been made from raw milk for hundreds of years. Food poisoning due to contamination of unpasteurised cheese in France is common.

Germany

In Germany, raw milk is sold as *Vorzugsmilch*. This means, the raw milk has to be packed before vending, with the necessary information (Producer, durability etc.) written on the product. The distribution license has stringent quality restrictions, and so just 80 farmers in Germany have one.

Unpacked raw milk must

- be bought at the farm itself

- be milk from that farm

- be from the day of or the day before production

- have a warning label "Raw Milk - boil before usage"

Packed raw milk is sold widely in all health food stores, large supermarkets, gourmet delis and delicatessen sections of department stores. Raw milk is legally sold in the entire country, and the same goes for raw milk cheeses, which are especially sought out and promoted by the health food and slow food movements.

Scandinavia

Shops are not permitted to sell unpasteurised milk to consumers in Norway Sweden Finland and Denmark. All four countries allow limited "barn door" sales subject to strict controls. One distributor in Denmark is licensed to supply restaurants with raw milk from approved farms. Pasteurisation of milk became common practice in Denmark and Sweden in the mid-1880s.

United Kingdom

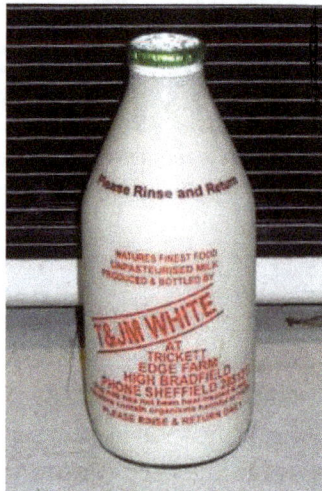

A bottle of green-top milk

Sales of raw drinking milk are prohibited in Scotland following a spate of deaths in 1983. While it is legal in England, Wales, and Northern Ireland, the only registered producers are in England. About 200 producers sell raw, or "green top" milk direct to consumers, either at the farm, at a farmers' market, or through a delivery service. The bottle must display the warning "this product has not been heat-treated and may contain organisms harmful to health", and the dairy must conform to higher hygiene standards than dairies producing only pasteurised milk.

As it is only legal to supply unpasteurised milk direct to consumers, it is illegal to be sold in the urban retail centers (high streets), via shops or supermarkets.

North America

Canada

The sale of raw milk directly to consumers is prohibited in Canada under the *Food and Drug Regulations* since 1991.

No person shall sell the normal lacteal secretion obtained from the mammary gland of the cow, genus Bos, or of any other animal, or sell a dairy product made with any such secretion, unless the secretion or dairy product has been pasteurized by being held at a temperature and for a period that ensure the reduction of the alkaline phosphatase activity so as to meet the tolerances specified in official method MFO-3, Determination of Phosphatase Activity in Dairy Products, dated November 30, 1981.

— Section B.08.002.2 (1)

Provincial laws also forbid the sale and distribution of raw milk. For instance, Ontario's Health Protection and Promotion Act, subsection 18(1) reads: "No person shall sell, offer for sale, deliver or distribute milk or cream that has not been pasteurized or sterilized in a plant that is licensed under the Milk Act or in a plant outside Ontario that meets the standards for plants licensed under the Milk Act."

In January 2010, Michael Schmidt was found not guilty on 19 charges relating to the sale of raw milk in the Ontario Court of Justice. On appeal to the Ontario Court of Justice, that decision was overturned. Schmidt was convicted on thirteen counts and imposed fines totaling $9,150 and one year of probation. A subsequent appeal to the Ontario Court of Appeal was dismissed.

In British Columbia, Alice Jongerden and Michael Schmidt and Gordon Watson — persons involved in the operation of her raw milk dairy — attempted to avoid enforcement of a judgement against them under the Public Health Act by challenging the constitutionality of the legislation, which deems raw milk to be a hazardous product, on the grounds that it violated the Canadian Charter of Rights and Freedoms. This argument, and other defenses invoked by her and defendants in her business, was rejected in 2013 by the Supreme Court of British Columbia, which instead found Schmidt and Watson guilty of civil contempt, and sentenced them to a 3 month suspended sentence imprisonment with a probationary period of 1 year during which "Any repetition of this contempt ... will trigger the imposed sentence imprisonment of 3 months." They were also charged special costs.

Meanwhile, Canada does permit the sale of raw milk cheeses that are aged over 60 days. In 2009, the province of Quebec modified regulations to allow raw milk cheeses aged less than 60 days provided stringent safeguards are met.

United States

In the early 20th century many states allowed the sale of raw milk that was certified by a "medical milk commission", effectively allowing an alternative of extra inspection for pasteurization. Now most states impose restrictions on raw milk suppliers due to concerns about safety. Twenty-eight U.S. states allow sales of raw milk. Cow shares can be found, and raw milk purchased for animal consumption in many states where retail for human consumption is prohibited. The sale of raw milk cheese is permitted as long as the cheese has been aged for 60 days or more.

The FDA reports that, in 2002, consuming partially heated raw milk and raw milk products caused 200 Americans to become ill in some manner.

Many governmental officials and the majority of public health organizations hold to the need for pasteurization. Before pasteurization, many dairies, especially in cities, fed their cattle on low-quality food, and their milk was rife with dangerous bacteria. Pasteurizing it was the only way to make it safely drinkable. As pasteurization has been standard for many years, it is now widely assumed that raw milk is dangerous. The Cornell University Food Science Department has compiled data indicating that pathogenic microorganisms are present in between 0.87% and 12.6% of raw milk samples.

Proponents of raw milk (in the U.S.) advance two basic arguments for unpasteurized milk. They state that pasteurization destroys or damages some of the milk's nutrients, and that while pasteurization may kill dangerous bacteria, it also kills off good bacteria that raw milk supporters have stated to have health benefits. The United States Food and Drug administration has stated that this is false, and that pasteurizing milk does not destroy any of its nutritive value.

Proponents also invoke the benefits of direct-marketing when promoting the sale of raw milk. The ability of the farmer to eliminate the middle-man and sell directly to the consumer allows for greater profitability. Many manufacturers sell small-scale pasteurization equipment, thereby allowing farmers to both bypass the milk processors and sell pasteurized milk directly to the consumer. Additionally, some small U.S. dairies are now beginning to adopt low-temperature vat pasteurization. Advocates of low-temperature vat pasteurization note that it produces a product similar to raw milk in composition and is not homogenized.

Alongside the ongoing empirical debate, food freedom advocates cite libertarian arguments in claiming a basic civil right of each person to weigh the risks and benefits in choosing the food one eats.

Oceania

Australia

The sale of raw milk for drinking purposes is illegal in all states and territories in Australia, as is all raw milk cheese. This has been circumvented somewhat by selling raw milk as *bath milk*. An exception to the cheese rule has been made recently for two Roquefort cheeses. There is some indication of share owning cows, allowing the "owners" to consume the raw milk, but also evidence that the government is trying to close this loophole.

On November 8, 2015, 4 year old Apu Khangura died of haemolytic uraemic syndrome, 7 other children became seriously ill, the Victorian government created new regulations which require producers to treat raw milk to reduce pathogens, or to make the product unpalatable to taste, such as making it bitter.

New Zealand

Raw milk products can be made and sold in New Zealand, but is highly regulated to offset the pathogen risk. Raw milk for drinking can only be sold directly from a producer (the farm gate) and only in amounts suitable for personal consumption (up to 5 litres).

Use

Raw yak milk is allowed to ferment overnight to become yak butter. Some cheeses are produced with raw milk although local statutes vary regarding what if any health precautions must be followed such as aging cheese for a certain amount of time.

A thick mixture known as Syllabub was created by milkmaids squirting milk directly from a cow into a container of cider, beer, or other beverage.

Milking Pipeline

A milking pipeline or milk pipeline is a component of a dairy farm animal-milking operation which is used to transfer milk from the animals to a cooling and storage bulk tank.

Setup

In small dairy farms with less than 100 cows, goats or sheep, the pipeline is installed above the animals' stalls and they are then are milked in sequence by moving down the row of stalls. The milking machine is a lightweight transportable hose assembly which is plugged into sealed access ports along the pipeline.

In the United States, for farmers who participate in the voluntary Dairy Herd Improvement Association, approximately once a month the milk volume from each animal is measured using additional portable metering devices inserted between the milker and the pipeline.

In large dairy farms with more than 100 animals, the pipeline is installed within a milking parlor that the animals walk through in order to be milked at fixed stations. Because the machine is stationary, it can include additional fixed equipment such as computerized milk-metering systems to measure volume, which would be cumbersome to use with portable milkers.

In both cases the pipeline is constructed out of stainless steel, which does not easily corrode and is resistant to most chemicals, though larger operations may use larger-diameter pipes in order to handle greater milk volumes.

Transfer from Pipeline to Bulk Tank

There is usually a transition point to move the milk from the pipeline under vacuum to the bulk tank, which is at normal atmospheric pressure. This is done by having the milk flow into a receiver bowl or globe, which is a large hollow glass container with electronic liquid-detecting probes in the center. As the milk rises to a certain height in the bowl, a transfer pump is used to push it through a one-way check valve and into a pipe that transfers it to the bulk tank. When the level has dropped far enough in the bowl, the transfer pump turns off. Without the check valve, the milk in the bulk tank could be sucked back into the receiver bowl when the pump is not running.

In the event of electronics or pump failure, there is also usually a secondary bowl attached to the top of receiver bowl, which contains a float and a diaphragm valve. If the main receiver bowl overflows due to pump failure, the rising milk lifts the float in the secondary bowl, which will cut off vacuum to the entire milk pipeline and will prevent the milk or wash water from being sucked into the vacuum pump.

Some milk handling systems eliminate the receiver bowl and transfer pump by having rubber seals on the bulk tank covers, to permit the entire tank to be under vacuum until milking is finished. Milk can then just flow directly by gravity from the pipeline into the bulk tank.

Pipeline Cleaning

320x240, 170 kilobit video of the pipeline cleaning process for a small 35-cow dairy farm, that has a traditional stanchion barn with haymow. The automatic washing system shown is a 1970s Bender Machine Works "Trol-O-Matic 5570", and the pipeline receiver and pump were made by Sta-Rite.

The pipeline and all milk handling systems are cleaned after every milking session using a washing system that first rinses out the remaining milk and then flushes cleaning solution through the piping to kill bacteria and remove *milkstone*, a layer of scale mainly formed by cations like calcium and magnesium. The entire washing mechanism is operated very much like a household dishwasher with an automatic fill system, soap dispenser, and automatic drain opener.

The pipeline is usually set up so that the vacuum in the system that lifts milk up can also be used to drive the cleaning process. Rather than having a single line run to the bulk tank, typically a pair of lines transport milk by gravity flow to the receiver bowl and transfer pump. The high ends of these two lines are joined together to form a complete loop back to the receiver bowl.

Cleaning is accomplished by inserting a choke plug into one of the lines leading to the transfer pump, and sucking large volumes of water from a wash-water supply tank into the choked line. This choke plug is mounted on a rod, and is inserted into the line before cleaning, and pulled out for regular milking. Due to the choke, the water, which is sufficient to completely fill the pipe, is sucked up one side of the pipeline, over the high point joining the two pipeline sections, and then flows back to the receiver bowl and transfer pump through the unchoked line. The transfer pump is then used to move the cleaning solution from the receiver bowl back to the wash-water supply tank to restart the process.

Typically, the inlet ports on the receiver globe are designed so that large slugs of wash water moving at high speed will enter on a tangent to the sides of the globe and rapidly spin around inside to assist in vigorous cleaning of the globe's interior. It is normal for wash water to overflow out the top of the globe and for some wash water to be sucked into the overflow chamber to also flush it out. During cleaning the bottom of the overflow chamber is connected to a drain channel on the receiver globe to permit water to flow out.

For the small-farm pipeline, portable milkers are inserted into this cleaning loop usually by sucking the cleaning solution out of the wash supply tank through the milker claw and outputting from the milker hoses into the choked end of the line. When the water returns to the receiver bowl, the transfer pump returns the water back to the milker's water pickup tank.

Automatic Milking

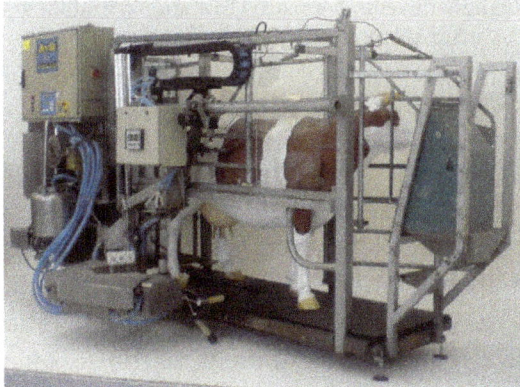

A Fullwood *Merlin* AMS unit from the 1990s, exhibit at the Deutsches Museum in Germany

Automatic milking is the milking of dairy animals, especially of dairy cattle, without human labour. Automatic milking systems (AMS), also called voluntary milking systems (VMS), were developed in the late 20th century. They have been commercially available since the early 1990s. The core of such systems that allows complete automation of the milking process is a type of agricultural robot. Automated milking is therefore also called robotic milking. Common systems rely on the use of computers and special herd management software.

Automation in Milking

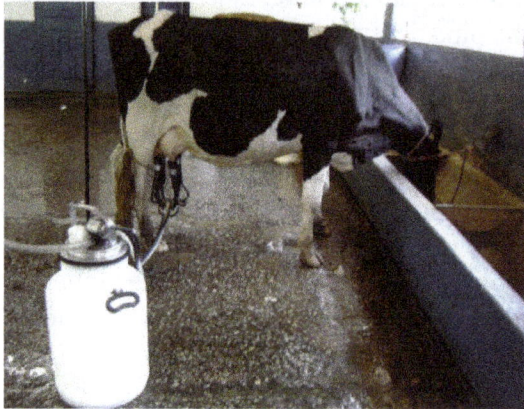

A cow and a milking machine – partial automation compared to hand milking

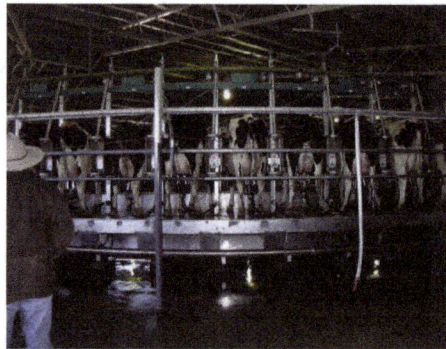

A rotary milking parlor – higher efficiency compared to stationary milking parlors, but still requiring manual labour with milking machines etc.

Basics – Milking Process and Milking Schedules

The milking process is the collection of tasks specifically devoted to extracting milk from an animal (rather than the broader field of dairy animal husbandry). This process may be broken down into several sub-tasks: collecting animals before milking, routing animals into the parlour, inspection and cleaning of teats, attachment of milking equipment to teats, and often massaging the back of the udder to relieve any held back milk, extraction of milk, removal of milking equipment, routing of animals out of the parlour.

Maintaining milk yield during the lactation period (approximately 300 days) requires consistent milking intervals, usually twice daily and with maximum time spacing between milkings. In fact all activities must be scheduled around the milking process on the dairy farm. Such a milking routine imposes restrictions on time management and personal life of an individual farmer, as the farmer is committed to milking in the early morning and in the evening for seven days a week regardless of personal health, family responsibilities or social schedule. This time restriction is exacerbated for lone farmers and farm families if extra labour cannot easily or economically be obtained, and is a factor in the decline in small-scale dairy farming. Techniques such as once-a-day milking and voluntary milking have been investigated to reduce these time constraints.

Automation Progress in the 20th Century

To alleviate the labour involved in milking, much of the milking process has been automated during the 20th century: many farmers use semi-automatic or automatic cow traffic control (powered gates, etc.), the milking machine (a basic form was developed in the late 19th century) has entirely automated milk extraction, and automatic cluster removal is available to remove milking equipment after milking. Automatic teat spraying systems are available, however there is some debate over the cleaning effectiveness of these.

The final manual labour tasks remaining in the milking process were cleaning and inspection of teats and attachment of milking equipment (milking cups) to teats. Automatic cleaning and attachment of milking cups is a complex task, requiring accurate detection of teat position and a dextrous mechanical manipulator. These tasks have been automated successfully in the voluntary milking system (VMS), or automatic milking system (AMS).

Automatic Milking Systems (AMS)

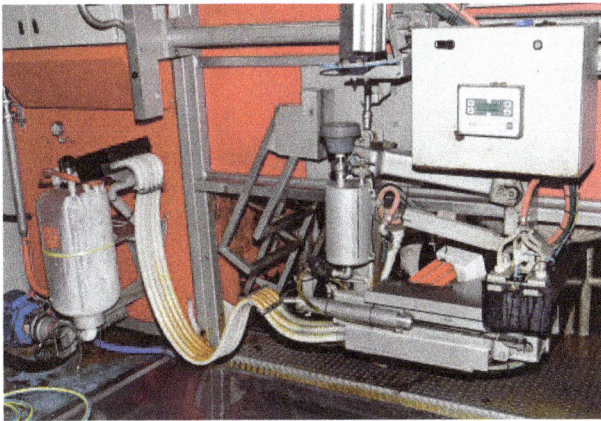

An older Lely *Astronaut* AMS unit at work (milking)

Since the 1970s, much research effort has been expended in investigating methods to alleviate time management constraints in conventional dairy farming, culminating in the development of the automated voluntary milking system. There is a video of the historical development of the milking robot at Silsoe Research Institute.

Voluntary milking allows the cow to decide her own milking time and interval, rather than being milked as part of a group at set milking times. AMS requires complete automation of the milking process as the cow may elect to be milked at any time during a 24-hour period.

The milking unit comprises a milking machine, a teat position sensor (usually a laser), a robotic arm for automatic teat-cup application and removal, and a gate system for controlling cow traffic. The cows may be permanently housed in a barn, and spend most of their time resting or feeding in the free-stall area. If cows are to be grazed as well, a selection gate is required to allow only those cows that have been milked to the outside pastures.

When the cow elects to enter the milking unit (due to highly palatable feed that she finds in the milking box), a cow ID sensor reads an identification tag (transponder) on the cow and passes the cow ID to the control system. If the cow has been milked too recently, the automatic gate system sends the cow out of the unit. If the cow may be milked, automatic teat cleaning, milking cup application, milking, and teatspraying takes place. As an incentive to attend the milking unit, concentrated feedstuffs needs to be fed to the cow in the milking unit.

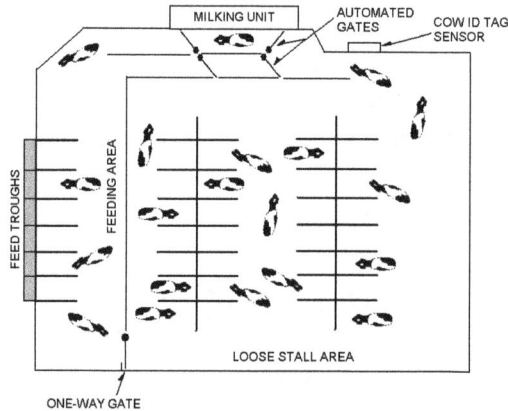

Typical VMS stall layout (*forced cow traffic* layout)

The barn may be arranged such that access to the main feeding area can only be obtained by passing the milking unit. This layout is referred to as *forced cow traffic*. Alternatively, the barn may be set up such that the cow always has access to feed, water, and a comfortable place to lie down, and is only motivated to visit the milking system by the palatable feed available there. This is referred to as *free cow traffic*.

The innovative core of the AMS system is the robotic manipulator in the milking unit. This robotic arm automates the tasks of teat cleaning and milking attachment and removes the final elements of manual labour from the milking process. Careful design of the robot arm and associated sensors and controls allows robust unsupervised performance, such that the farmer is only required to attend the cows for condition inspection and when a cow has not attended for milking.

Typical capacity for an AMS is 50–70 cows per milking unit. AMS usually achieve milking frequencies between 2 and 3 times per day, so a single milking unit handling 60 cows and milking each cow 3 times per day has a capacity of 7.5 cows per hour. This low capacity is convenient for lower-cost design of the robot arm and associated control system, as a window of several minutes is available for each cow and high-speed operation is not required.

AMS units have been available commercially since the early 1990s, and have proved relatively successful in implementing the voluntary milking method. Many of the research and developments have taken place in the Netherlands. The most farms with AMS are located in the Netherlands, and Denmark.

A new variation on the theme of robotic milking includes a similar robotic arm system, but coupled with a rotary platform, improving the number of cows that can be handled per robot arm.

Advantages

An AMS unit at work (teat cleaning)

- Elimination of labour - The farmer is freed from the milking process and associated rigid schedule, and labour is devoted to supervision of animals, feeding, etc.

- Milking consistency – The milking process is consistent for every cow and every visit, and is not influenced by different persons milking the cows. The four separate milking cups are removed individually, meaning that an empty quarter does not stay attached while the other three are finishing, resulting in less threat of injury. The newest models of automatic milkers can vary the pulsation rate and vacuum level based on milk flow from each quarter.

- Increased milking frequency – Milking frequency may increase to three times per day, however typically 2.5 times per day is achieved. This may result in less stress on the udder and increased comfort for the cow, as on average less milk is stored. Higher frequency milking increases milk yield per cow, however much of this increase is water rather than solids.

- Perceived lower stress environment – There is a perception that elective milking schedules reduce cow stress. A study found no decrease in stress between automatic and conventional milking.

- Herd management – The use of computer control allows greater scope for data collection. Such data allows the farmer to improve management through analysis of trends in the herd, for example response of milk production to changes in feedstuffs. Individual cow histories may also be examined, and alerts set to warn the farmer of unusual changes indicating illness or injury. Information gathering provides added value for AMS, however correct interpretation and use of such information is highly dependent on the skills of the user or the accuracy of computer algorithms to create attention reports.

Considerations and Disadvantages

- Higher initial cost – AMS systems cost approximately €120,000 ($190,524) per milking unit as of 2003 (presuming barn space is already available for loose-stall housing). Equipment costs decreased from $175,000 for the first stall to $158,000. Equipment costs decreased from $10,000/stall for a double-six parlor to $9000/stall for a double-ten parlor with a cost of $1200/stall for pipeline milking. Initial parlor cost was increased $5000/stall to represent a high cost parlor. Whether it is economically beneficial to invest in an AMS instead of a conventional milking parlor depends on constructions costs, investments in the milking system and costs of labour. Besides costs of labour, the availability of labour should also be taken into account. In general, an AMS is economically beneficial for smaller scale farms, and large dairies can usually operate more cheaply with a milking parlor.

- Increased electricity costs - to operate the robots, but this can be more than outweighed by reduced labour input.

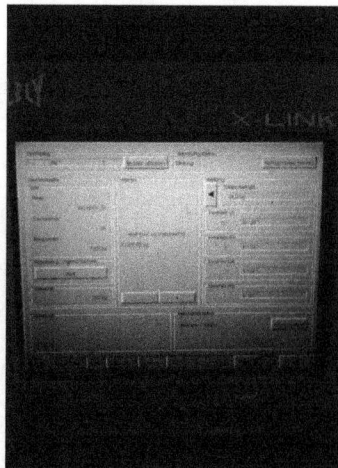

Touchscreen display of a milking robot

- Increased complexity – While complexity of equipment is a necessary part of technological advancement, the increased complexity of the AMS milking unit over conventional systems, increases the reliance on manufacturer maintenance services and possibly increasing operating costs. The farmer is exposed in the event of total system failure, relying on prompt response from the service provider. In practice AMS systems have proved robust and manufacturers provide good service networks. Because all milking cows have to visit the AMS voluntarily, the system requires a high quality of management. The system also involves a central place for the computer in the daily working routines.

- Difficult to apply in pasture systems – AMS works best in zero-grazing systems, in which the cow is housed indoors for most of the lactation period. Zero-grazing suits areas (e.g. the Netherlands) where land is at a premium, as maximum land can be devoted to feed production which is then collected by the farmer and brought to the animals in the barn. In pasture systems, cows graze in fields and are required to walk to the milking parlour. It has been found that cows tend not to attend the milking unit if the distance to walk is too great. There are currently research projects at the Dexcel facility in New Zealand, University of Sydney's FutureDairy site, and Michigan State University's Kellogg Biological Station, where cattle are on pasture and milked by AMS.

- Lower milk quality – Somatic cell count (SCC) and Plate loop count (PLC) are, respectively, measurements of the quantity of white blood cells and total number of bacteria present in a milk sample. A high SCC indicates reduced udder health (as the immune system fights some infection) and implies lower milk quality. AMS herds consistently show higher SCCs than conventionally milked herds. A high PLC indicates bacterial contamination, usually through poor sanitation or cooling and similarly implies low milk quality. High PLC in AMS may be attributed to the continuous use of milking lines (rather than twice a day in conventional systems), which reduces the time window for cleaning, and the incremental addition of milk to the bulk milk tank which may not cool efficiently at low milk levels.

- Possible increase in stress for some cows – Cows are social animals, and it has been found that due to dominance of some cows, others will be forced to milk only at night. Such behaviour is inconsistent with the perception that AM reduces stress by allowing "free choice" of milking time.

- Decreased contact between farmer and herd – Effective animal husbandry requires that the farmer be fully aware of herd condition. In conventional milking, the cows are observed before milking equipment is attached, and ill or injured cows can be earmarked for attention. Automatic milking removes the farmer from such close contact with the animal, with the possibility that illness may go unnoticed for longer periods and both milk quality and cow welfare suffer.

In practice, milk quality sensors at the milking unit attempt to detect changes in milk due to infection, and farmers inspect the herd frequently. However this concern has meant that farmers are still tied to a seven-day schedule. Modern automatic milking systems attempt to rectify this problem by gathering data that would not be available in many conventional systems including milk temperature, milk conductivity, milk color including infrared scan, change in milking speed, change in milking time or milk letdown by quarter, cow's weight, cow's activity (movements), time spent ruminating, etc.

Manufacturers

A DeLaval *VMS* unit, 2007

- Lely (Netherlands), *Lely Astronaut AMS*

- DeLaval (Sweden), *DeLaval VMS*

- Fullwood (UK), *Merlin AMS*

- GEA Farm Technologies (Germany, formerly WestfaliaSurge), *MIone AMS*

- SAC (Denmark), purchased the Dutch manufacturer of the *Galaxy Robot AMS* in 2005, sell under the brands *SAC RDS Futureline MARK II, Insentec Galaxy Starline, BouMatic's ProFlex*

- BoumaticRobotics (NL), *MR-S1, MR-D1*

References

- Blood DC, Studdert VP, Gay CC (2007). Saunders Comprehensive Veterinary Dictionary. St. Louis, Missouri, USA: Saunders Elsevierv. ISBN 0-7020-2789-8.

- McGee, Harold (2004) [1984]. "Milk and Dairy Products". On Food and Cooking: The Science and Lore of the Kitchen (2nd ed.). New York: Scribner. pp. 7–67. ISBN 978-0-684-80001-1.

- Bellwood, Peter (2005). "The Beginnings of Agriculture in Southwest Asia". First Farmers: the origins of agricultural societies. Malden, MA: Blackwell Publushing. pp. 44–68. ISBN 978-0-631-20566-1.

- Bellwood, Peter (2005). "Early Agriculture in the Americas". First Farmers: the origins of agricultural societies. Malden, MA: Blackwell Publushing. pp. 146–179. ISBN 978-0-631-20566-1.

- Price, T. D. (2000). "Europe's first farmers: an introduction". In T. D. Price. Europe's First Farmers. Cambridge: Cambridge University Press. pp. 1–18. ISBN 0-521-66203-6.

- Anthony, D. W. (2007). The Horse, the Wheel, and Language. Princeton, NJ: Princeton University Press. ISBN 978-0-691-05887-0.

- Gifford-Gonzalez, D. (2004). "Pastoralism and its Consequences". In A. B. Stahl. African archaeology: a critical introduction. Malden, MA: Blackwell Publishing. pp. 187–224. ISBN 978-1-4051-0155-4.

- Valenze, D. M. (2011). "Virtuous White Liquor in the Middle Ages". Milk: a local and global history. New Haven: Yale University Press. p. 34. ISBN 9780300117240.

- Carlisle, Rodney (2004). Scientific American Inventions and Discoveries, p.357. John Wiley & Songs, Inc., new Jersey. ISBN 0-471-24410-4.

- Hwang, Andy; Huang, Lihan (January 31, 2009). Ready-to-Eat Foods: Microbial Concerns and Control Measures. CRC Press. p. 88. ISBN 978-1-4200-6862-7. Retrieved April 19, 2011.

- Handbook of Food and Beverage Fermentation Technology. 2004. p. 265. ISBN 0-203-91355-8. Retrieved 6 September 2016.

- Designing Foods: Animal Product Options in the Marketplace. National Academies Press. 1988. ISBN 978-0-309-03795-2.

Powdered Milk Manufacturing and Processes

Powdered milk is a dairy product that is dried milk. One of the main purposes of drying milk is by preserving it. Some of the methods are spray drying, drum drying and freeze-drying. The aspects elucidated in this chapter are of vital importance and provides a better understanding of powdered milk manufacturing.

Powdered Milk

Powdered milk or dried milk is a manufactured dairy product made by evaporating milk to dryness. One purpose of drying milk is to preserve it; milk powder has a far longer shelf life than liquid milk and does not need to be refrigerated, due to its low moisture content. Another purpose is to reduce its bulk for economy of transportation. Powdered milk and dairy products include such items as dry whole milk, nonfat (skimmed) dry milk, dry buttermilk, dry whey products and dry dairy blends. Many dairy products exported conform to standards laid out in Codex Alimentarius. Many forms of milk powder are traded on exchanges.

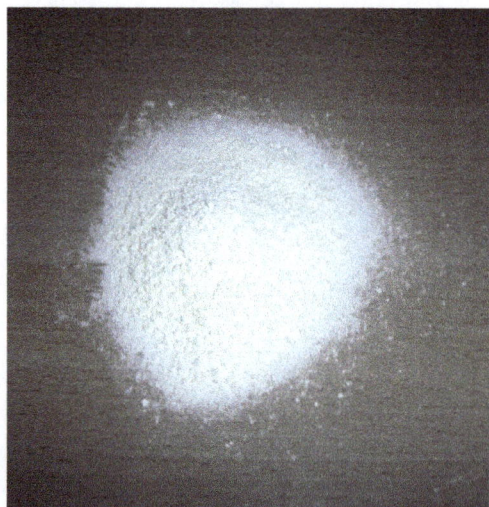

Powdered milk

Powdered milk is used for food and health (nutrition), and also in biotechnology (saturating agent).

History and Manufacture

Modified dry whole milk, fortified with vitamin D. This is the original container from 1947, provided by the Ministry of Food in London, England

While Marco Polo wrote of Mongolian Tatar troops in the time of Kublai Khan who carried sun-dried skimmed milk as "a kind of paste", the first modern production process for dried milk was invented by the Russian physician Osip Krichevsky in 1802. The first commercial production of dried milk was organized by the Russian chemist M. Dirchoff in 1832. In 1855, T.S. Grimwade took a patent on a dried milk procedure, though a William Newton had patented a vacuum drying process as early as 1837.

In modern times, powdered milk is usually made by spray drying nonfat skimmed milk, whole milk, buttermilk or whey. Pasteurized milk is first concentrated in an evaporator to approximately 50% milk solids. The resulting concentrated milk is then sprayed into a heated chamber where the water almost instantly evaporates, leaving fine particles of powdered milk solids.

Alternatively, the milk can be dried by drum drying. Milk is applied as a thin film to the surface of a heated drum, and the dried milk solids are then scraped off. However, powdered milk made this way tends to have a cooked flavor, due to caramelization caused by greater heat exposure.

Another process is freeze drying, which preserves many nutrients in milk, compared to drum drying.

The drying method and the heat treatment of the milk as it is processed alters the properties of the milk powder, such as its solubility in cold water, its flavor, and its bulk density.

Food and Health Uses

Incolac powdered milk

Powdered milk is frequently used in the manufacture of infant formula, confectionery such as chocolate and caramel candy, and in recipes for baked goods where adding liquid milk would render the product too thin. Powdered milk is also widely used in various sweets such as the famous Indian milk balls known as gulab jamun and a popular Indian sweet delicacy (sprinkled with desiccated coconut) known as Chum chum (made with skim milk powder).

Powdered milk is also a common item in UN food aid supplies, fallout shelters, warehouses, and wherever fresh milk is not a viable option. It is widely used in many developing countries because of reduced transport and storage costs (reduced bulk and weight, no refrigerated vehicles). Like other dry foods, it is considered nonperishable, and is favored by survivalists, hikers, and others requiring nonperishable, easy-to-prepare food.

Because of its resemblance to cocaine and other drugs, powdered milk is sometimes used in filmmaking as a non-toxic prop that may be insufflated.

Reconstitution

The weight of nonfat dry milk (NFDM) to use is about 10% of the water weight. Alternatively, one cup of potable fluid milk from powdered milk requires one cup of potable water and one-third cup of powdered milk.

Nutritional Value

Milk powders contain all twenty-one standard amino acids, the building blocks of proteins, and are high in soluble vitamins and minerals. According to USAID, the typical average amounts of major nutrients in the unreconstituted nonfat dry milk are (by weight) 36% protein, 52% carbohydrates (predominantly lactose), calcium 1.3%, potassium 1.8%. Whole milk powder, on the other hand, contains on average 25-27% protein, 36-38% carbohydrates, 26-40% fat, and 5-7% ash (minerals). However, inappropriate storage conditions such as high relative humidity and high ambient temperature can significantly degrade the nutritive value of milk powder.

Commercial milk powders are reported to contain oxysterols (oxidized cholesterol) in higher amounts than in fresh milk (up to 30 µg/g, versus trace amounts in fresh milk). Oxysterols are derivatives of cholesterol that are produced either by free radicals or by enzymes. Some free radicals-derived oxysterols have been suspected of being initiators of atherosclerotic plaques. For comparison, powdered eggs contain even more oxysterols, up to 200 µg/g.

Export Market

National household dried machine skimmed milk. This was U.S.-produced dry milk for food export in June 1944.

European production of milk powder is estimated around 800,000 tons of which the main volume is exported in bulk packing or consumer packs.

Brands on the market include "Nido", from the company Nestlé, "Incolac" from the company Belgomilk, and "Dutch Lady" from FrieslandCampina.

Adulteration

In the 2008 Chinese milk scandal, melamine adulterant was found in Sanlu infant formula, added to fool tests into reporting higher protein content. Thousands became ill, and some children died, after consuming the product.

In August 2013, China temporarily suspended all milk powder imports from New Zealand, after a scare where botulism-causing bacteria was falsely detected in several batches of New Zealand-produced whey protein concentrate. As a result of the product recall, the New Zealand dollar slipped significantly based on expected losses in sales from this single commodity.

Use in Biotechnology

Fat-free powdered milk is used as a saturating agent to block nonspecific binding sites on supports like blotting membranes (nitrocellulose, polyvinylidene fluoride (PVDF) or nylon), preventing binding of further detection reagents and subsequent background. It may be referred as Blotto. The major protein of milk, casein, is responsible for most of the binding site saturation effect.

Spray Drying

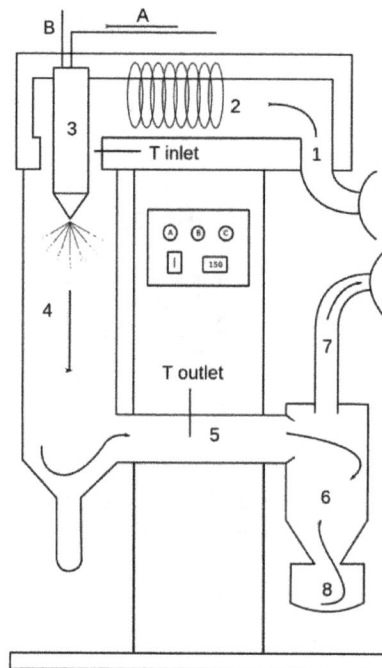

Laboratory-scale spray dryer.
A=Solution or suspension to be dried in, B=Atomization gas in,
1= Drying gas in, 2=Heating of drying gas, 3=Spraying of solution or suspension,
4=Drying chamber, 5=Part between drying chamber and cyclone, 6=Cyclone, 7=Drying gas is taken
away, 8=Collection vessel of product, arrows mean that this is co-current lab-spraydryer

Spray drying is a method of producing a dry powder from a liquid or slurry by rapidly drying with a hot gas. This is the preferred method of drying of many thermally-sensitive materials such as foods and pharmaceuticals. A consistent particle size distribution is a reason for spray drying some industrial products such as catalysts. Air is the heated drying medium; however, if the liquid is a flammable solvent such as ethanol or the product is oxygen-sensitive then nitrogen is used.

All spray dryers use some type of atomizer or spray nozzle to disperse the liquid or slurry into a controlled drop size spray. The most common of these are rotary disk and single-fluid high pressure swirl nozzles. Atomizer wheels are known to provide broader particle size distribution, but both methods allow for consistent distribution of particle size. Alternatively, for some applications two-fluid or ultrasonic nozzles are used. Depending on the process needs, drop sizes from 10 to 500 μm can be achieved with the appropriate choices. The most common applications are in the 100 to 200 μm diameter range. The dry powder is often free-flowing.

The most common type of spray dryers are called single effect. There is a single source of drying air at the top of the chamber. In most cases the air is blown in the same direction as the sprayed liquid (co-current). A fine powder is produced, but it can have poor flow and produce a lot of dust. To overcome the dust and poor flow of the powder, a new generation of spray dryers called multiple effect spray dryers have been produced. Instead of drying the liquid in one stage, drying is done through two steps: the first at the top (as per single effect) and the second with an integrated static bed at the bottom of the chamber. The bed provides a humid environment which causes smaller particles to clump, producing more uniform particle sizes, usually within the range of 100 to 300 μm. These powders are free-flowing due to the larger particle size.

The fine powders generated by the first stage drying can be recycled in continuous flow either at the top of the chamber (around the sprayed liquid) or at the bottom, inside the integrated fluidized bed. The drying of the powder can be finalized on an external vibrating fluidized bed.

The hot drying gas can be passed in as a co-current, same direction as sprayed liquid atomizer, or counter-current, where the hot air flows against the flow from the atomizer. With co-current flow, particles spend less time in the system and the particle separator (typically a cyclone device). With counter-current flow, particles spend more time in the system and is usually paired with a fluidized bed system. Co-current flow generally allows the system to operate more efficiently.

Alternatives to spray dryers are:

1. Freeze dryer: a more-expensive batch process for products that degrade in spray drying. Dry product is not free-flowing.

2. Drum dryer: a less-expensive continuous process for low-value products; creates flakes instead of free-flowing powder.

3. Pulse combustion dryer: A less-expensive continuous process that can handle higher viscosities and solids loading than a spray dryer, and that sometimes gives a freeze-dry quality powder that is free-flowing.

Spray Dryer

A spray dryer takes a liquid stream and separates the solute or suspension as a solid and the solvent into a vapor. The solid is usually collected in a drum or cyclone. The liquid input stream is sprayed through a nozzle into a hot vapor stream and vaporized. Solids form as moisture quickly leaves the droplets. A nozzle is usually used to make the droplets as small as possible, maximizing heat transfer and the rate of water vaporization. Droplet sizes can range from 20 to 180 μm depending on the nozzle. There are two main types of nozzles: high pressure single fluid nozzle (50 to 300 bars) and two-fluid nozzles: one fluid is the liquid to dry and the second is compressed gas (generally air at 1 to 7 bars).

Spray drying nozzles.

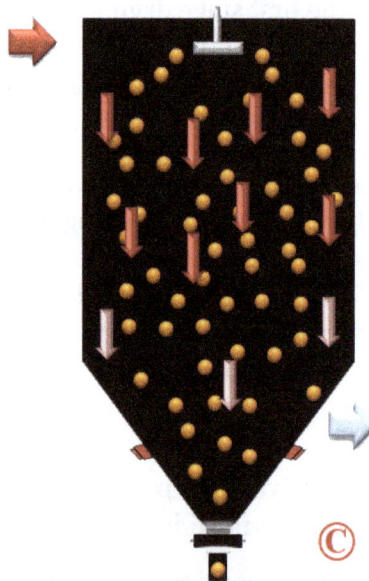

Schematic illustration of spray drying process.

Spray dryers can dry a product very quickly compared to other methods of drying. They also turn a solution, or slurry into a dried powder in a single step, which can be advantageous as it simplifies the process and improves profit margins.

Micro-encapsulation

Spray drying often is used as an encapsulation technique by the food and other industries. A substance to be encapsulated (the load) and an amphipathic carrier (usually some sort of modified starch) are homogenized as a suspension in water (the slurry). The slurry is then fed into a spray drier, usually a tower heated to temperatures well over the boiling point of water.

As the slurry enters the tower, it is atomized. Partly because of the high surface tension of water and partly because of the hydrophobic/hydrophilic interactions between the amphipathic carrier, the water, and the load, the atomized slurry forms micelles. The small size of the drops (averaging 100 micrometers in diameter) results in a relatively large surface area which dries quickly. As the water dries, the carrier forms a hardened shell around the load.

Load loss is usually a function of molecular weight. That is, lighter molecules tend to boil off in larger quantities at the processing temperatures. Loss is minimized industrially by spraying into taller towers. A larger volume of air has a lower average humidity as the process proceeds. By the osmosis principle, water will be encouraged by its difference in fugacities in the vapor and liquid phases to leave the micelles and enter the air. Therefore, the same percentage of water can be dried out of the particles at lower temperatures if larger towers are used. Alternatively, the slurry can be sprayed into a partial vacuum. Since the boiling point of a solvent is the temperature at which the vapor pressure of the solvent is equal to the ambient pressure, reducing pressure in the tower has the effect of lowering the boiling point of the solvent.

The application of the spray drying encapsulation technique is to prepare "dehydrated" powders of substances which do not have any water to dehydrate. For example, instant drink mixes are spray dries of the various chemicals which make up the beverage. The technique was once used to remove water from food products; for instance, in the preparation of dehydrated milk. Because the milk was not being encapsulated and because spray drying causes thermal degradation, milk dehydration and similar processes have been replaced by other dehydration techniques. Skim milk powders are still widely produced using spray drying technology around the world, typically at high solids concentration for maximum drying efficiency. Thermal degradation of products can be overcome by using lower operating temperatures and larger chamber sizes for increased residence times.

Recent research is now suggesting that the use of spray-drying techniques may be an alternative method for crystallization of amorphous powders during the drying process since the temperature effects on the amorphous powders may be significant depending on drying residence times.

Spray Drying Applications

Food: milk powder, coffee, tea, eggs, cereal, spices, flavorings, blood, starch and starch derivatives, vitamins, enzymes, stevia, nutracutical, colourings, etc.

Pharmaceutical: antibiotics, medical ingredients, additives

Industrial: paint pigments, ceramic materials, catalyst supports, microalgae

Nano Spray Dryer

The nano spray dryer offers new possibilities in the field of spray drying. Particles can be produced in the range of 300 nm to 5 μm with a narrow size distribution. High yields, up to 90%, can be produced and the minimal sample amount is 1 mL.

Drum Drying

Drum drying is a method used for drying out liquids from raw materials with drying drum. In the drum-drying process, pureed raw ingredients are dried at relatively low temperatures over rotating, high-capacity drums that produce sheets of drum-dried product. This product is milled to a finished flake or powder form. Modern drum drying techniques results in dried ingredients which reconstitute immediately and retain much of their original flavor, color and nutritional value.

Some advantages of drum drying include the ability of drum dryers to dry viscous foods which cannot be easily dried with other methods. Drum dryers can be clean and hygienic and easy to operate and maintain.

Other products where drum drying can be used are, for example, starches, breakfast cereals, baby food, instant mashed potatoes to make them cold-water-soluble.

Freeze-drying

Freeze dried ice cream

Freeze-drying—technically known as lyophilisation, lyophilization, or cryodesiccation—is a dehydration process typically used to preserve a perishable material or make the material more convenient for transport. Freeze-drying works by freezing the material and then reducing the surrounding pressure to allow the frozen water in the material to sublimate directly from the solid phase to the gas phase.

A benchtop manifold freeze-drier

History

The process of freeze-drying was invented in 1906 by Arsène d'Arsonval and his assistant Frédéric Bordas at the laboratory of biophysics of Collège de France in Paris. In 1911 Downey Harris and Shackle developed the freeze-drying method of preserving live rabies virus which eventually led to development of the first antirabies vaccine.

Modern freeze-drying was developed during World War II. Blood serum being sent to Europe from the US for medical treatment of the wounded required refrigeration, but because of the lack of simultaneous refrigeration and transport, many serum supplies were spoiling before reaching their intended recipients. The freeze-drying process was developed as a commercial technique that enabled serum to be rendered chemically stable and viable without having to be refrigerated. Shortly thereafter, the freeze-dry process was applied to penicillin and bone, and lyophilization became recognized as an important technique for preservation of biologicals. Since that time, freeze-drying has been used as a preservation or processing technique for a wide variety of products. These applications include the following but are not limited to: the processing of food, pharmaceuticals, and diagnostic kits; the restoration of water damaged documents; the preparation of river-bottom sludge for hydrocarbon analysis; the manufacturing of ceramics used in the semiconductor industry; the production of synthetic skin; the manufacture of sulfur-coated vials; and the restoration of historic/reclaimed boat hulls.

Stages

There are four stages in the complete drying process: pretreatment, freezing, primary drying, and secondary drying.

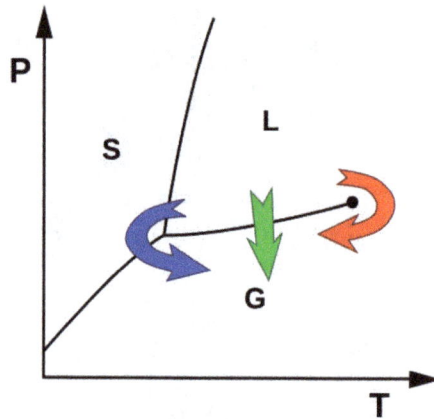

In a typical phase diagram, the boundary between gas and liquid runs from the triple point to the critical point. Freeze-drying (blue arrow) brings the system around the triple point, avoiding the direct liquid-gas transition seen in ordinary drying time (green arrow).

Pretreatment

Pretreatment includes any method of treating the product prior to freezing. This may include concentrating the product, formulation revision (i.e., addition of components to increase stability, preserve appearance, and/or improve processing), decreasing a high-vapor-pressure solvent, or increasing the surface area. In many instances the decision to pretreat a product is based on theoretical knowledge of freeze-drying and its requirements, or is demanded by cycle time or product quality considerations.

Freezing

In a lab, this is often done by placing the material in a freeze-drying flask and rotating the flask in a bath, called a shell freezer, which is cooled by mechanical refrigeration, dry ice in aqueous methanol, or liquid nitrogen. On a larger scale, freezing is usually done using a freeze-drying machine. In this step, it is important to cool the material below its triple point, the lowest temperature at which the solid and liquid phases of the material can coexist. This ensures that sublimation rather than melting will occur in the following steps. Larger crystals are easier to freeze-dry. To produce larger crystals, the product should be frozen slowly or can be cycled up and down in temperature. This cycling process is called annealing. However, in the case of food, or objects with formerly-living cells, large ice crystals will break the cell walls (a problem discovered, and solved, by Clarence Birdseye), resulting in the destruction of more cells, which can result in increasingly poor texture and nutritive content. In this case, the freezing is done rapidly, in order to lower the material to below its eutectic point quickly, thus avoiding the formation of ice crystals. Usually, the freezing temperatures are between −50 °C and −80 °C (-58 °F and -112 °F) . The freezing phase is the most critical in the whole freeze-drying process, because the product can be spoiled if improperly done.

Amorphous materials do not have a eutectic point, but they do have a critical point, below which the product must be maintained to prevent melt-back or collapse during primary and secondary drying.

Primary Drying

During the primary drying phase, the pressure is lowered (to the range of a few milli-bars), and enough heat is supplied to the material for the ice to sublime. The amount of heat necessary can be calculated using the sublimating molecules' latent heat of subli-mation. In this initial drying phase, about 95% of the water in the material is sublimat-ed. This phase may be slow (can be several days in the industry), because, if too much heat is added, the material's structure could be altered.

In this phase, pressure is controlled through the application of partial vacuum. The vacuum speeds up the sublimation, making it useful as a deliberate drying process. Furthermore, a cold condenser chamber and/or condenser plates provide a surface(s) for the water vapour to re-solidify on. This condenser plays no role in keeping the ma-terial frozen; rather, it prevents water vapor from reaching the vacuum pump, which could degrade the pump's performance. Condenser temperatures are typically below −50 °C (−58 °F).

It is important to note that, in this range of pressure, the heat is brought mainly by conduction or radiation; the convection effect is negligible, due to the low air density.

Secondary Drying

The secondary drying phase aims to remove unfrozen water molecules, since the ice was removed in the primary drying phase. This part of the freeze-drying process is gov-erned by the material's adsorption isotherms. In this phase, the temperature is raised higher than in the primary drying phase, and can even be above 0 °C, to break any physico-chemical interactions that have formed between the water molecules and the frozen material. Usually the pressure is also lowered in this stage to encourage desorp-tion (typically in the range of microbars, or fractions of a pascal). However, there are products that benefit from increased pressure as well.

After the freeze-drying process is complete, the vacuum is usually broken with an inert gas, such as nitrogen, before the material is sealed.

At the end of the operation, the final residual water content in the product is extremely low, around 1% to 4%.

Properties of Freeze-dried Products

If a freeze-dried substance is sealed to prevent the reabsorption of moisture, the sub-stance may be stored at room temperature without refrigeration, and be protected against spoilage for many years. Preservation is possible because the greatly reduced water content inhibits the action of microorganisms and enzymes that would normally spoil or degrade the substance.

Freeze-drying also causes less damage to the substance than other dehydration methods using higher temperatures. Freeze-drying does not usually cause shrinkage or toughening of the material being dried. In addition, flavours, smells and nutritional content generally remain unchanged, making the process popular for preserving food. However, water is not the only chemical capable of sublimation, and the loss of other volatile compounds such as acetic acid (vinegar) and alcohols can yield undesirable results.

Freeze-dried products can be rehydrated (reconstituted) much more quickly and easily because the process leaves microscopic pores. The pores are created by the ice crystals that sublimate, leaving gaps or pores in their place. This is especially important when it comes to pharmaceutical uses. Freeze-drying can also be used to increase the shelf life of some pharmaceuticals for many years.

Protectants

Similar to cryoprotectants, some molecules protect freeze-dried material. Known as lyoprotectants, these molecules are typically polyhydroxy compounds such as sugars (mono-, di-, and polysaccharides), polyalcohols, and their derivatives. Trehalose and sucrose are natural lyoprotectants. Trehalose is produced by a variety of plant (for example selaginella and arabidopsis thaliana), fungi, and invertebrate animals that remain in a state of suspended animation during periods of drought (also known as anhydrobiosis).

Applications

Pharmaceutical and Biotechnology

Lyophilized 5% w/v sucrose cake in a pharmaceutical glass vial

Pharmaceutical companies often use freeze-drying to increase the shelf life of the products, such as live virus vaccines, biologics and other injectables. By removing the water from the material and sealing the material in a glass vial, the material can be easily stored, shipped, and later reconstituted to its original form for injection. Another example from the pharmaceutical industry is the use of freeze drying to produce tablets or wafers, the advantage of which is less excipient as well as a rapidly absorbed and easily administered dosage form.

Freeze-dried pharmaceutical products are produced as lyophilized powders for reconstitution in vials and more recently in prefilled syringes for self-administration by a patient. Many biopharmaceutical products based on therapeutic proteins such as monoclonal antibodies require lyophilization for stability. Examples of lyophilized biopharmaceuticals include blockbuster drugs such as Etanercept (Enbrel by Pfizer), Infliximab (Remicade by Janssen Biotech), Rituximab and Trastuzumab (Herceptin by Genentech).

Freeze-drying is also used in manufacturing of raw materials for pharmaceutical products. Active Pharmaceutical Product Ingredients (APIs) are lyophilized to achieve chemical stability under room temperature storage. Bulk lyophilization of APIs is typically conducted using trays instead of glass vials.

Dry powders of probiotics are often produced by bulk freeze-drying of live microorganisms such as Lactic acid bacteria and Bifidobacteria.

Food and Agriculture-based Industries

Freeze dried bacon bars

Freeze-dried coffee, a form of instant coffee

Although freeze-drying is used to preserve food, its earliest use in agriculturally based industries was in processing of crops such as peanuts/groundnuts and tobacco in the early 1970s. Because heat, commonly used in crop and food processing, invariably alters the structure and chemistry of the product, the main objective of freeze-drying is to avoid heat and thus preserve the structural and chemical integrity/composition with little or no alteration. Therefore, freeze-dried crops and foods are closest to the natural composition with respect to structure and chemistry. The process came to wide public attention when it was used to create freeze-dried ice cream, an example of astronaut food. It is also widely used to produce essences or flavourings to add to food.

Because of its light weight per volume of reconstituted food, freeze-dried products are popular and convenient for hikers. More dried food can be carried per the same weight of wet food, and remains in good condition for longer than wet food, which tends to spoil quickly. Hikers reconstitute the food with water available at point of use.

Instant coffee is sometimes freeze-dried, despite the high costs of the freeze-driers used. The coffee is often dried by vaporization in a hot air flow, or by projection onto hot metallic plates. Freeze-dried fruits are used in some breakfast cereal or sold as a snack, and are an especially popular snack choice among toddlers, preschoolers, hikers and dieters, as well as being used by some pet owners as a treat for pet birds. Most commercial freezing is done either in cold air kept in motion by fans (blast freezing) or by placing the food-stuffs in packages or metal trays on refrigerated surfaces (contact freezing).

Culinary herbs, vegetables (such as vitamin-rich spinach and watercress), the temperature sensitive baker`s yeast suspension and the nutrient-rich pre-boiled rice can also be freeze-dried. During three hours of drying the spinach and watercress has lost over 98% of its water content, followed by the yeast suspension with 96% and the pre-boiled rice by 75%. The air-dried herbs are far more common and less expensive. Freeze dried tofu is a popular foodstuff in Japan ("Koya-dofu" or "shimi-dofu" in Japanese).

Technological Industry

In chemical synthesis, products are often freeze-dried to make them more stable, or easier to dissolve in water for subsequent use.

In bioseparations, freeze-drying can be used also as a late-stage purification procedure, because it can effectively remove solvents. Furthermore, it is capable of concentrating substances with low molecular weights that are too small to be removed by a filtration membrane. Freeze-drying is a relatively expensive process. The equipment is about three times as expensive as the equipment used for other separation processes, and the high energy demands lead to high energy costs. Furthermore, freeze-drying also has a long process time, because the addition of too much heat to the material can cause melting or structural deformations. Therefore, freeze-drying is often reserved for materials that are heat-sensitive, such as proteins, enzymes, microorganisms, and blood plasma. The low operating temperature of the process leads to minimal damage of these heat-sensitive products.

In nanotechnology, freeze-drying is used for nanotube purification to avoid aggregation due to capillary forces during regular thermal vaporization drying.

Other Uses

Organizations such as the Document Conservation Laboratory at the United States National Archives and Records Administration (NARA) have done studies on freeze-drying as a recovery method of water-damaged books and documents. While recovery is possible, restoration quality depends on the material of the documents. If a document is made of a variety of materials, which have different absorption properties, expansion will occur at a non-uniform rate, which could lead to deformations. Water can also cause mold to grow or make inks bleed. In these cases, freeze-drying may not be an effective restoration method.

In bacteriology freeze-drying is used to conserve special strains.

In high-altitude environments, the low temperatures and pressures can sometimes produce natural mummies by a process of freeze-drying.

Advanced ceramics processes sometimes use freeze-drying to create a formable powder from a sprayed slurry mist. Freeze-drying creates softer particles with a more homogeneous chemical composition than traditional hot spray drying, but it is also more expensive.

Freeze-drying is also used for floral preservation. Wedding bouquet preservation has become very popular with brides who want to preserve their wedding day flowers

A new form of burial which previously freeze-dries the body with liquid nitrogen has been developed by the Swedish company Promessa Organic AB, which puts it forward as an environmentally friendly alternative to traditional casket and cremation burials.

Equipment

There are essentially three categories of freeze-dryers: the manifold freeze-dryer, the rotary freeze-dryer and the tray style freeze-dryer. Two components are common to all types of freeze-dryers: a vacuum pump to reduce the ambient gas pressure in a vessel containing the substance to be dried and a condenser to remove the moisture by condensation on a surface cooled to −40 to −80 °C (−40 to −112 °F). The manifold, rotary and tray type freeze-dryers differ in the method by which the dried substance is interfaced with a condenser. In manifold freeze-dryers a short usually circular tube is used to connect multiple containers with the dried product to a condenser. The rotary and tray freeze-dryers have a single large reservoir for the dried substance.

Unloading trays of freeze-dried material from a small cabinet-type freeze-dryer

Rotary freeze-dryers are usually used for drying pellets, cubes and other pourable substances. The rotary dryers have a cylindrical reservoir that is rotated during drying to achieve a more uniform drying throughout the substance. Tray style freeze-dryers usually have rectangular reservoir with shelves on which products, such as pharmaceutical solutions and tissue extracts, can be placed in trays, vials and other containers.

Manifold freeze-dryers are usually used in a laboratory setting when drying liquid substances in small containers and when the product will be used in a short period of time. A manifold dryer will dry the product to less than 5% moisture content. Without heat, only primary drying (removal of the unbound water) can be achieved. A heater must be added for secondary drying, which will remove the bound water and will produce a lower moisture content.

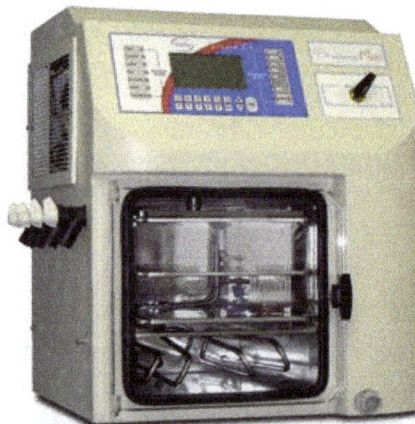

Product viewable single shelf freeze-dryer

Tray style freeze-dryers are typically larger than the manifold dryers and are more sophisticated. Tray style freeze-dryers are used to dry a variety of materials. A tray freeze-dryer is used to produce the driest product for long-term storage. A tray freeze-dryer allows the product to be frozen in place and performs both primary (unbound water removal) and secondary (bound water removal) freeze-drying, thus producing the driest possible end-product. Tray freeze-dryers can dry products in bulk or in vials or other containers. When drying in vials, the freeze-dryer is supplied with a stoppering mechanism that allows a stopper to be pressed into place, sealing the vial before it is exposed to the atmosphere. This is used for long-term storage, such as vaccines.

Improved freeze-drying techniques are being developed to extend the range of products that can be freeze-dried, to improve the quality of the product, and to produce the product faster with less labor.

Production freeze-drier

References

- Wayne Gisslen (2009). Professional Baking (5th ed.). Hoboken, N.J.: John Wiley. p. 77. ISBN 0-471-78349-8. Retrieved May 18, 2011.

- p 655 in "Advanced Dairy Chemistry: Volume 2 - Lipids" by P.F. Fox and P. McSweeney, Birkhäuser, 2006 ISBN 978-0-387-26364-9

- p 296 in "Toxins in Food" by W.M. Dabrowski and Z.E. Sikorski, CRC Press, 2004, ISBN 978-0-8493-1904-4

- p 82 in "Lab Ref, Volume 2: A Handbook of Recipes, Reagents, and Other Reference Tools for Use at the Bench" by A.S. Mellick and L. Rodgers, CSHL Press, 2002, ISBN 978-0-87969-630-6

- Encyclopedia of Agriculture, Food, and Biological Engineering. Marcel Dekker, Inc. pp. 211–214. ISBN 0-8247-0938-1.

- В Рязани производят кошерное сухое молоко [Kosher Milk is Produced in Ryazan]. Наша Рязань (in Russian). 16 February 2010. Retrieved 22 April 2013.

- Brian Donaldson. "Wedding Bouquet Preservation...saver your special memories of your wedding day through the preservation of your bridal bouquet!" (MHT). Retrieved 2010-06-23.

5

Yogurt and Curd: A Comprehensive Study

Yogurt is a food product that is produced from milk. The other dairy products discussed are strained yogurt, doogh, curd and buffalo curd. The topics discussed in the section help the readers in broadening the existing knowledge on yogurt and curd.

Yogurt

Yogurt, yoghurt, or yoghourt is a food produced by bacterial fermentation of milk.

The bacteria used to make yogurt are known as "yogurt cultures". Fermentation of lactose by these bacteria produces lactic acid, which acts on milk protein to give yogurt its texture and characteristic tang. Cow's milk is commonly available worldwide, and, as such, is the milk most commonly used to make yogurt. Milk from water buffalo, goats, ewes, mares, camels, and yaks is also used to produce yogurt where available locally. Milk used may be homogenized or not (milk distributed in many parts of the world is homogenized); both types may be used, with substantially different results.

Yogurt is produced using a culture of *Lactobacillus delbrueckii* subsp. *bulgaricus* and *Streptococcus thermophilus* bacteria. In addition, other lactobacilli and bifidobacteria are also sometimes added during or after culturing yogurt. Some countries require yogurt to contain a certain amount of colony-forming units of bacteria; in China, for example, the requirement for the number of lactobacillus bacteria is at least 1×10^6 CFU per gram per milliliter. To produce yogurt, milk is first heated, usually to about 85 °C (185 °F), to denature the milk proteins so that they set together rather than form curds. After heating, the milk is allowed to cool to about 45 °C (113 °F). The bacterial culture is mixed in, and a temperature of 45 °C (113 °F) is maintained for four to seven hours to allow fermentation.

Etymology and Spelling

The word is derived from Turkish: *yoğurt*, and is usually related to the verb *yoğurmak*, "to knead", or "to be curdled or coagulated; to thicken". It may be related to *yoğun*, meaning thick or dense. The letter ğ was traditionally rendered as "gh" in transliterations of Turkish prior to 1928. In older Turkish, the letter denoted a voiced velar fricative, but this sound is elided between back vowels in modern Turkish, in which the word is pronounced.

In English, there are several variations of the spelling of the word, including *yogurt*, *yoghurt* and to a lesser extent *yoghourt*, *yogourt*, *yaghourt*, *yahourth*, *yoghurd*, *joghourt*, and *jogourt*. In the United Kingdom and Australia, *yogurt* and *yoghurt* are both current, *yogurt* being used by the Australian and British dairy councils, and *yoghourt* is an uncommon alternative. In the United States, Canada, and New Zealand, *yogurt* is the usual spelling and *yoghurt* a minor variant.

Historically, there have been cases of yogurt being spelled with a "J" instead of a "Y" (e.g. *jogurt* and *joghurt*) due to alternative transliteration methods. However, there has been a decline in these variations in English speaking countries; in certain European countries, on the other hand, it is still commonly spelled with a "J". Most people tend to spell in the manner shown on the packaging of the major brands in their country.

Whatever the spelling, the word is usually pronounced with a short in England and Wales, and with a long in Scotland, North America, Australia, New Zealand, Ireland and South Africa.

History

Analysis of the *L. delbrueckii* subsp. *bulgaricus* genome indicates that the bacterium may have originated on the surface of a plant. Milk may have become spontaneously and unintentionally exposed to it through contact with plants, or bacteria may have been transferred from the udder of domestic milk-producing animals.

The origins of yogurt are unknown, but it is thought to have been invented in Mesopotamia around 5000 BC.

In ancient Indian records, the combination of yogurt and honey is called "the food of the gods". Persian traditions hold that "Abraham owed his fecundity and longevity to the regular ingestion of yogurt".

Unstirred Turkish *Süzme Yoğurt* (strained yogurt), with a 10% fat content

The cuisine of ancient Greece included a dairy product known as oxygala which is believed to have been a form of yogurt. Galen (AD 129 – c. 200/c. 216) mentioned that oxygala was consumed with honey, similar to the way thickened Greek yogurt is eaten today.

The oldest writings mentioning yogurt are attributed to Pliny the Elder, who remarked that certain "barbarous nations" knew how "to thicken the milk into a substance with an agreeable acidity". The use of yogurt by medieval Turks is recorded in the books *Diwan Lughat al-Turk* by Mahmud Kashgari and *Kutadgu Bilig* by Yusuf Has Hajib written in the 11th century. Both texts mention the word "yogurt" in different sections and describe its use by nomadic Turks. The earliest yogurts were probably spontaneously fermented by wild bacteria in goat skin bags.

Some accounts suggest that Indian emperor Akbar's cooks would flavor yogurt with mustard seeds and cinnamon. Another early account of a European encounter with yogurt occurs in French clinical history: Francis I suffered from a severe diarrhea which no French doctor could cure. His ally Suleiman the Magnificent sent a doctor, who allegedly cured the patient with yogurt. Being grateful, the French king spread around the information about the food which had cured him.

Until the 1900s, yogurt was a staple in diets of people in the Russian Empire (and especially Central Asia and the Caucasus), Western Asia, South Eastern Europe/Balkans, Central Europe, and India. Stamen Grigorov (1878–1945), a Bulgarian student of medicine in Geneva, first examined the microflora of the Bulgarian yogurt. In 1905, he described it as consisting of a spherical and a rod-like lactic acid-producing bacteria. In 1907, the rod-like bacterium was called *Bacillus bulgaricus* (now *Lactobacillus delbrueckii subsp. bulgaricus*). The Russian Nobel laureate and biologist Ilya Ilyich Mechnikov (also known as Élie Metchnikoff), from the Institut Pasteur in Paris, was influenced by Grigorov's work and hypothesized that regular consumption of yogurt was responsible for the unusually long lifespans of Bulgarian peasants. Believing *Lactobacillus* to be essential for good health, Mechnikov worked to popularize yogurt as a foodstuff throughout Europe.

Isaac Carasso industrialized the production of yogurt. In 1919, Carasso, who was from Ottoman Salonika, started a small yogurt business in Barcelona, Spain, and named the business Danone ("little Daniel") after his son. The brand later expanded to the United States under an Americanized version of the name: Dannon.

Yogurt with added fruit jam was patented in 1933 by the Radlická Mlékárna dairy in Prague.

Yogurt was introduced to the United States in the first decade of the twentieth century, influenced by Élie Metchnikoff's *The Prolongation of Life; Optimistic Studies* (1908); it was available in tablet form for those with digestive intolerance and for home culturing. It was popularized by John Harvey Kellogg at the Battle Creek Sanitarium, where it was used both orally and in enemas, and later by Armenian immigrants Sarkis and Rose Colombosian, who started "Colombo and Sons Creamery" in Andover, Massachusetts in 1929. Colombo Yogurt was originally delivered around New England in a horse-drawn wagon inscribed with the Armenian word "madzoon" which was later changed to "yogurt", the Turkish name of the product, as Turkish was the lingua franca between immigrants of the various Near Eastern ethnicities who were the main consumers at that time. Yogurt's popularity in the United States was enhanced in the 1950s and 1960s, when it was presented as a health food. By the late 20th century, yogurt had become a common American food item and Colombo Yogurt was sold in 1993 to General Mills, which discontinued the brand in 2010.

Nutrition and Health

In a 100-gram amount providing 406 kilojoules (97 kcal) of dietary energy, yogurt (plain Greek yogurt from whole milk) is 81% water, 9% protein, 5% fat and 4% carbohydrates, including 4% sugars (table). As a proportion of the Daily Value (DV), a serving of yogurt is a rich source of vitamin B_{12} (31% DV) and riboflavin (23% DV), with moderate content of protein, phosphorus and selenium (14 to 19% DV; table).

Comparison of Whole Dairy Milk and Plain Yogurt from Whole Dairy Milk, One Cup (245 g) Each		
Property	**Milk**	**Yogurt**
kilo calories	146	149
Total Fat	7.9 g	8.5 g
Cholesterol	24.4 mg	11 mg
Sodium	98 mg	113 mg
Phosphorus	222 mg	233 mg
Potassium	349 mg	380 mg
Total Carbohydrates	12.8 g	12 g
Protein	7.9 g	9 g
Vitamin A	249 IU	243 IU
Vitamin C	0.0 mg	1.2 mg
Vitamin D	96.5 IU	~
Vitamin E	0.1 mg	0.1 mg
Vitamin K	0.5 µg	0.5 µg
Thiamine	0.1 mg	0.1 mg
Riboflavin	0.3 mg	0.3 mg
Niacin	0.3 mg	0.2 mg
Vitamin B6	0.1 mg	0.1 mg
Folate	12.2 µg	17.2 µg
Vitamin B12	1.1 µg	0.9 µg
Choline	34.9 mg	1.0 mg

Betaine	1.5 mg	~
Water	215 g	215 g
Ash	1.7 g	1.8 g

Tilde (~) represents missing or incomplete data. – The above shows that there is little difference between whole milk and yogurt made from whole milk with respect to the listed nutritional constituents. The differences may be explained as a result of testing the product after draining liquid whey from the yogurt thereby changing the percentage of that constituent in the final product.

Although yogurt is often associated with probiotics having positive effects on immune, cardiovascular or metabolic health, there is insufficient high-quality clinical evidence to conclude that consuming yogurt lowers risk of diseases or improves health.

Lactose-intolerant individuals may tolerate yogurt better than other dairy products due to the conversion of lactose to the sugars glucose and galactose, and the fermentation of lactose to lactic acid carried out by the bacteria present in the yogurt.

Varieties and Presentation

| *Tzatziki* is a side dish made with yogurt, popular in Greek cuisine, and similar yet thicker than the Turkish Cacik and close to the traditional Bulgarian Milk salad. | *Skyr* is an Icelandic cultured dairy product, similar to strained yogurt. It has been a part of Icelandic cuisine for over a thousand years. It is traditionally served cold with milk and a topping of sugar. | *Cacık*, a Turkish cold appetizer made from yogurt |

Da-hi is a yogurt of the Indian subcontinent, known for its characteristic taste and consistency. The word *da-hi* seems to be derived from the Sanskrit word *dadhi*, one of the five elixirs, or panchamrita, often used in Hindu ritual. *Dahi* also holds cultural symbolism in many homes in the *Mithila* region of Nepal and Bihar. Yogurt balances the palate across regional cuisines throughout India. In the hot and humid south, yogurt and foods made of yogurt are a staple in order to cool down – yogurt rice is always the last dish of the meal. Also, the vegetarian population of India derives some protein from yogurt (other than lentil and beans). Sweet yogurt (*meesti doi* or *meethi dahi*) is common in eastern parts of India, made by fermenting sweetened milk. While cow's milk is considered sacred and is currently the primary ingredient for yogurt, goat and buffalo milk were widely used in the past, and valued for the fat content. Butter and cream were made by churning the yogurt/milk.

In India and Pakistan, it is often used in cosmetics mixed with turmeric and honey. Sour yogurt, is also used as a hair conditioner by women in many parts of India and Pakistan. *Dahi* is also known as *Mosaru* (Kannada), *Thayir* (Tamil/Malayalam), *doi* (Assamese, Bengali), *dohi* (Odia), *perugu* (Telugu), *Qəzana a pəəner* (Pashto) and *Dhahi* or *Dhaunro*.

Raita is a yogurt-based South Asian/Indian condiment, used as a side dish. The yogurt is seasoned with coriander (cilantro), cumin, mint, cayenne pepper, and other herbs and spices. Vegetables such as cucumber and onions are mixed in, and the mixture is served chilled. Raita has a cooling effect on the palate which makes it a good foil for spicy Indian and Pakistani dishes. Raita is sometimes also referred to as *dahi*.

Dadiah or dadih is a traditional West Sumatran yogurt made from water buffalo milk, fermented in bamboo tubes.

Yogurt is popular in Nepal, where it is served as both an appetizer and dessert. Locally called *dahi*, it is a part of the Nepali culture, used in local festivals, marriage ceremonies, parties, religious occasions, family gatherings, and so on. The most famous type of Nepalese yogurt is called *juju dhau*, originating from the city of Bhaktapur.

In Tibet, yak milk (technically dri milk, as the word yak refers to the male animal) is made into yogurt (and butter and cheese) and consumed.

In Northern Iran, *Mâst Chekide* is a variety of kefir yogurt with a distinct sour taste. It is usually mixed with a pesto-like water and fresh herb purée called delal. Yogurt is a side dish to all Iranian meals. The most popular appetizers are spinach or eggplant borani, *Mâst-o-Khiâr* with cucumber, spring onions and herbs, and *Mâst-Musir* with wild shallots. In the summertime, yogurt and ice cubes are mixed together with cucumbers, raisins, salt, pepper and onions and topped with some croutons made of Persian traditional bread and served as a cold soup. Ashe-Mâst is a warm yogurt soup with fresh herbs, spinach and lentils. Even the leftover water extracted when straining yogurt is cooked to make a sour cream sauce called kashk, which is usually used as a topping on soups and stews.

Matsoni is a Georgian yogurt popular in the Caucasus and Russia. It is used in a wide variety of Georgian dishes and is believed to have contributed to the high life expectancy and longevity in the country. Dannon used this theory in their 1978 TV advertisement called *In Soviet Georgia* where shots of elderly Georgian farmers were interspersed with an off-camera announcer intoning, "In Soviet Georgia, where they eat a lot of yogurt, a lot of people live past 100." Matsoni is also popular in Japan under the name Caspian Sea Yogurt.

Tarator and Cacık are popular cold soups made from yogurt, popular during summertime in Albania, Azerbaijan (known as Dogramac), Bulgaria, Macedonia, Serbia and Turkey. They are made with ayran, cucumbers, dill, salt, olive oil, and optionally garlic and ground walnuts. Tzatziki in Greece and milk salad in Bulgaria are thick yogurt-based salads similar to tarator.

Khyar w Laban (cucumber and yogurt salad) is a popular dish in Lebanon and Syria. Also, a wide variety of local Lebanese and Syrian dishes are cooked with yogurt like "Kibbi bi Laban", etc.

Rahmjoghurt, a creamy yogurt with much higher fat content (10%) than many yogurts offered in English-speaking countries (*Rahm* is German for "cream"), is available in Germany and other countries.

Dovga, a yogurt soup cooked with a variety of herbs and rice is popular in Azerbaijan, often served warm in winter or refreshingly cold in summer.

Yogurt made with unhomogenized milk is sometimes called cream-top yogurt; a layer of cream rises to the top.

Jameed is yogurt which is salted and dried to preserve it. It is popular in Jordan.

Zabadi is the type of yogurt made in Egypt, usually from the milk of the Egyptian water buffalo. It is particularly associated with Ramadan fasting, as it is thought to prevent thirst during all-day fasting.

Sweetened and Flavored Yogurt

To offset its natural sourness, yogurt is also sold sweetened, flavored or in containers with fruit or fruit jam on the bottom. The two styles of yogurt commonly found in the grocery store are set type yogurt and Swiss style yogurt. Set type yogurt is when the yogurt is packaged with the fruit on the bottom of the cup and the yogurt on top. Swiss style yogurt is when the fruit is blended into the yogurt prior to packaging.

Lassi and Moru are common beverages in India. Lassi is stirred liquified curd that is either salted or sweetened with sugar commonly, less commonly honey and often combined with fruit pulp to create flavored lassi. Mango lassi is a western favorite, as is coconut lassi. Consistency can vary widely, with urban and commercial lassis being of uniform texture through being processed, whereas rural and rustic lassi has curds in it, and sometimes has malai (cream) added or removed. Moru is a popular South Indian summer drink, meant to keep drinkers hydrated through the hot and humid summers of the South. It is prepared by considerably thinning down yogurt with water, adding salt (for electrolyte balance) and spices, usually green chili peppers, asafoetida, curry leaves and mustard.

Large amounts of sugar – or other sweeteners for low-energy yogurts – are often used in commercial yogurt. Some yogurts contain added starch, pectin (found naturally in fruit), and/or gelatin to create thickness and creaminess artificially at lower cost. Gelatin is a meat or fish product, therefore vegetarians should avoid products containing it. This type of yogurt is also marketed under the name Swiss-style, although it is unrelated to the way yogurt is eaten in Switzerland. Some yogurts, often called "cream line", are made with whole milk which has not been homogenized so the cream rises to the top.

In the UK, Ireland, France and United States, sweetened, flavored yogurt is the most popular type, typically sold in single-serving plastic cups. Common flavors include vanilla, honey, and toffee, and fruit such as strawberry, cherry, blueberry, blackberry, raspberry, mango and peach. In the early twenty-first century yogurt flavors inspired by desserts, such as chocolate or cheesecake, have been available.

There is concern about the health effects of sweetened yogurt. The United Kingdom and the United States recommend different maximum amounts of daily sugar intake but in both nations many sweetened yogurts have too much.

A 150g (5oz) serving of some 0% fat yogurts can contain as much as 20g (0.7oz) of sugar – the equivalent of five teaspoons, says Action on Sugar – which is about 40% of a woman's daily recommended intake of added sugar (50g or 1.7oz) and about 30% of that for men (70g or 2.5oz).

The American Heart Association recommends that men eat no more than 36 grams of sugar per day, and women no more than 20. Many of the top-selling yogurts have even more than the 19 grams of sugar that is contained in a Twinkie.

Consumers wanting sweetened yogurt are advised to choose yogurt sweetened with sugar substitute and check the contents list to avoid corn syrup, high fructose corn syrup, honey, or sugar.

Strained Yogurt

Use coffee filter to strain yogurt in a home refrigerator.

Strained yogurt is yogurt which has been strained through a filter, traditionally made of muslin and more recently of paper or cloth. This removes the whey, giving a much thicker consistency and a distinctive slightly tangy taste. Strained yogurt is becoming more popular with those who make yogurt at home, especially if using skimmed milk which results in a thinner consistency.

Yogurt that has been strained to filter or remove the whey is known as Labneh in Middle Eastern countries. It has a consistency between that of yogurt and cheese. It is popular for sandwiches in Middle Eastern countries. Olive oil, cucumber slices, olives, and various green herbs may be added. It can be thickened further and rolled into balls, preserved in olive oil, and fermented for a few more weeks. It is sometimes used with onions, meat, and nuts as a stuffing for a variety of pies or kibbeh balls.

Some types of strained yogurts are boiled in open vats first, so that the liquid content is reduced. The popular East Indian dessert, a variation of traditional dahi called mishti dahi, offers a thicker, more custard-like consistency, and is usually sweeter than western yogurts.

Strained yogurt is also enjoyed in Greece and is the main component of *tzatziki* (from Turkish "cacık"), a well-known accompaniment to gyros and souvlaki pita sandwiches: it is a yogurt sauce or dip made with the addition of grated cucumber, olive oil, salt and, optionally, mashed garlic.

Srikhand, a popular dessert in India, is made from strained yogurt, saffron, cardamom, nutmeg and sugar and sometimes fruits such as mango or pineapple.

In North America and Britain, strained yogurt is commonly called "Greek yogurt". Strained yogurt is sometimes marketed in North America as "Greek yogurt" and in Britain as "Greek-style yoghurt". In Britain the name "Greek" may only be applied to yogurt made in Greece.

Beverages

PCC Dairy Yogurt Milk, with live cultures, made from water buffalo's cream milk Philippine Carabao Center.

Ayran

Doogh ("dawghe" in Neo-Aramaic), ayran or dhallë is a yogurt-based, salty drink popular in Iran, Albania, Bulgaria, Turkey, Azerbaijan, Afghanistan, Pakistan, Bangladesh, Macedonia, Uzbekistan, Kazakhstan and Kyrgyzstan. It is made by mixing yogurt with water and (sometimes) salt. The same drink is known *tan* in Armenia; *laban ayran* in Syria and Lebanon; *shenina* in Iraq and Jordan; *laban arbil* in Iraq; *majjiga* (Telugu), *majjige* (Kannada), and *moru* (Tamil and Malayalam) in South India; namkeen *lassi* in Punjab and all over Pakistan.

Borhani (or Burhani) is a spicy yogurt drink popular in Bangladesh and parts of Bengal. It is usually served with kacchi biryani at weddings and special feasts. Key ingredients are yogurt blended with mint leaves (mentha), mustard seeds and black rock salt (Kala Namak). Ground roasted cumin, ground white pepper, green chili pepper paste and sugar are often added.

Lassi is a yogurt-based beverage originally from the Indian subcontinent that is usually slightly salty or sweet. Lassi is a staple in Punjab. In some parts of the subcontinent, the sweet version may be commercially flavored with rosewater, mango or other fruit juice to create a very different drink. Salty lassi is usually flavored with ground, roasted cumin and red chilies; this salty variation may also use buttermilk, and in India is interchangeably called *ghol* (Bengal), *mattha* (North India), "majjige" (Karnataka), *majjiga* (Telangana & Andhra Pradesh), *moru* (Tamil Nadu and Kerala), *Dahi paani Chalha* (Odisha), *tak* (Maharashtra), or *chaas* (Gujarat). Lassi is very widely drunk in India, Pakistan, and Bangladesh. Mango Lassi is a popular drink at Indian restaurants in US.

In Bosnia and Herzegovina, Croatia, Macedonia, Montenegro, Serbia, and Slovenia, an unsweetened and unsalted yogurt drink usually called simply *jogurt* is a popular accompaniment to *burek* and other baked goods.

Sweetened yogurt drinks are the usual form in Europe (including the UK) and the US, containing fruit and added sweeteners. These are typically called "drinkable yogurt".

Also available are "yogurt smoothies", which contain a higher proportion of fruit and are more like smoothies. In Ecuador, yogurt smoothies flavored with native fruit are served with pan de yuca as a common type of fast food.

Also in Turkey, yogurt soup or *Yayla Çorbası* is a popular way of consuming yogurt. The soup is a mix of yogurt, rice, flour and dried mint.

Plant-milk Yogurt

Plant-milk yogurt

A variety of plant-milk yogurts appeared in the 2000s, using soy milk, rice milk, and nut milks such as almond milk and coconut milk. So far the most widely sold variety of plant milk yogurts is soy yogurt. These yogurts are suitable for vegans, people with intolerance to dairy milk, and those who prefer plant milks.

Making Yogurt at Home

Commercially available yogurt maker

Yogurt is made by heating milk to a temperature that denaturates its proteins (scalding), essential for making yogurt, cooling it to a temperature that will not kill the live microorganisms that turn the milk into yogurt, inoculating certain bacteria (starter culture), usually *Streptococcus thermophilus* and *Lactobacillus bulgaricus*, into the milk, and finally keeping it warm for several hours. The milk may be held at 85 °C (185 °F) for a few minutes, or boiled (giving a somewhat different result). It must be cooled to 50 °C (122 °F) or somewhat less, typically 40–46 °C (104–115 °F). Starter culture must then be mixed in well, and the mixture must be kept undisturbed and warm for several hours, ranging from 5 to 12, with longer fermentation producing a more acid yogurt. The starter culture may be a small amount of live yogurt; dried starter culture is available commercially.

Home yogurt maker

Milk with a higher concentration of solids than normal milk may be used; the higher solids content produces a firmer yogurt. Solids can be increased by adding dried milk.

The yogurt-making process provides two significant barriers to pathogen growth, heat and acidity (low pH). Both are necessary to ensure a safe product. Acidity alone has been questioned by recent outbreaks of food poisoning by *E. coli O157:H7* that is acid-tolerant. *E. coli O157:H7* is easily destroyed by pasteurization (heating); the initial heating of the milk kills pathogens as well as denaturing proteins. The microorganisms that turn milk into yogurt can tolerate higher temperatures than most pathogens, so that a suitable temperature not only encourages the formation of yogurt, but inhibits pathogenic microorganisms.

Once the yogurt has formed it can, if desired, be strained to reduce the whey content and thicken it.

Strained Yogurt

Strained yogurt, Greek yogurt, yogurt cheese, or labneh is yogurt that has been strained to remove its whey, resulting in a thicker consistency than unstrained yogurt, while preserving yogurt's distinctive, sour taste. Like many types of yogurt, strained yogurt is often made from milk that has been enriched by boiling off some of its water content, or by adding extra butterfat and powdered milk. In Europe and North America, it is often made with low-fat or fat-free yogurt. In Scandinavia, particularly Iceland, a similar product, skyr is produced.

Strained yogurt is sometimes marketed in North America as "Greek yogurt" and in Britain as "Greek-style yoghurt", though strained yogurt is also widely eaten in Levantine, Eastern Mediterranean, Near Eastern, Central Asian and South Asian cuisines, wherein it is often used in cooking (as it is high enough in fat content to avoid curdling at higher temperatures). Such dishes may be cooked or raw, savory or sweet. Due to the straining process to remove excess whey, even non-fat varieties of strained yogurt are much thicker, richer, and creamier than yogurts that have not been strained.

In western Europe and the US, strained yogurt has increased in popularity compared to unstrained yogurt. Since the straining process removes some of the lactose, strained yogurt is lower in sugar than unstrained yogurt.

It was reported in 2012 that most of the growth in the $4.1 billion US yogurt industry came from the strained yogurt sub-segment, sometimes marketed as "Greek yogurt". In the US there is no legal definition of Greek yogurt, and yogurt thickened with thickening agents may also be sold as "Greek yogurt".

Nutrition

Strained yogurt contains a higher protein density than regular yogurt. The protein in strained yogurt is largely casein protein.

Variations

Yogurt being strained through a cheesecloth

Armenia

In Armenia, strained yogurt is called *kamats matzoon*. Traditionally, it was produced for long-term preservation by draining matzoon in cloth sacks. Afterwards it was stored in leather sacks or clay pots for a month or more depending on the degree of salting.

Bosnia and Herzegovina, Croatia, Serbia, and Macedonia

In the countries of the former Yugoslavia, strained yogurt made of cow's milk has become very popular in recent years. It is usually labeled *grčki tip jogurta* and eaten on its own as a snack or dessert.

Bulgaria

In Bulgaria, where yogurt is considered to be an integral part of the national cuisine, strained yogurt is called "tsedeno kiselo mlyako", and is used in a variety of salads and dressings. Another similar product is "katak"which is often made from sheep or goat milk.

Central Asia

In the cuisines of many Iranian, Baloch, and Turkic peoples (e.g. in Azerbaijani, Afghan, Tatar, Tajik, Uzbek, and other Central Asian cuisines), a type of strained yogurt called *chak(k)a* or *suzma* is consumed. It is obtained by draining qatiq, a local yogurt variety. By further drying it, one obtains qurut, a kind of dry fresh cheese.

Cyprus

As in Greece, strained yogurt is widely used in Cypriot cuisine both as an ingredient in recipes as well as on its own or as a supplement to a dish. In Cyprus, strained yogurt is usually made from sheep's milk.

Greece

Strained yogurt is used in Greek food mostly as the base for tzatziki dip and as a dessert, with honey, sour cherry syrup, or spoon sweets often served on top. A few savoury Greek dishes use strained yogurt. In Greece, strained yogurt, like yogurt in general, is traditionally made from sheep's milk. Fage International S.A. began straining cows milk yogurt for industrial production in Greece in 1975, which is when it launched its brand "Total".

Indian Subcontinent

In the Indian subcontinent, regular unstrained yogurt (*curd*), made from cow or water buffalo milk, it is often sold in disposable clay bowls called kulhar. Kept for a couple of hours in its clay pot, some of the water evaporates through the unfired clay's pores. It also cools the curd due to evaporation. But true strained yogurt is made by draining the yoghurt in a cloth.

A disposable clay pot with "dahi"

Shrikhand is an Indian dessert (often eaten with poori) made with strained yogurt and sugar, saffron, cardamom, diced fruit and nuts mixed in. It is particularly popular in the state of Gujarat and Maharashtra, where dairy producers market shrikhand similar to ice cream. In Pashtun-dominated regions of Pakistan and Afghanistan, a strained yogurt known as *chaka* is often consumed with rice and meat dishes.

Middle East

Labneh (also known as Labni, Lebni) is popular in the Levant and the Arabian Peninsula. Besides being used fresh, labneh is also dried then formed into balls, sometimes covered with herbs or spices, and stored under olive oil. Labneh is a popular mezze dish and sandwich ingredient. A common sandwich in the Middle East is one of labneh, mint, thyme, and olive on pita bread. The flavour depends largely on the sort of milk used: labneh from cow's milk has a rather mild flavour. Also the quality of olive oil topping influences the taste of labneh. Milk from camels and other animals is used in labneh production in Saudi Arabia and other Gulf countries.

Bedouin also produce a dry, hard labneh (*labaneh malboudeh*, similar to Central Asian qurut) that can be stored. Strained labneh is pressed in cheese cloth between two heavy stones and later sun dried. This dry labneh is often eaten with khubz (Arabic bread), in which both khubz and labneh are mixed with water, animal fat, and salt, and rolled into balls.

In Iraq, Jordan, Lebanon, Israel and Syria, Labneh is made by straining the liquid out of yogurt until it takes on a consistency similar to a soft cheese. It tastes like tart sour cream or heavy strained yogurt and is a common breakfast dip. It is usually eaten in a fashion similar to hummus, spread on a plate and drizzled with olive oil and often, dried mint.

Labneh is also the main ingredient in jameed, which is in turn used in mansaf, the national dish of Jordan.

Labneh is commonly consumed by both Jewish and Palestinian Israelis, often with pita and za'atar, or dried hyssop. It can also be purchased in the form of small white balls immersed in olive oil.

Labneh is a common breakfast food among Palestinians in the West Bank and the Gaza Strip.

Labaneh bil zayit, "labaneh in oil", consists of small balls of dry labneh kept under oil, where it can be preserved for over a year. As it ages it turns more sour.

Strained yogurt in Iran is called *mâst chekide* and is usually used for making dips, or served as a side dish. In Northern Iran, *mâst chekide* is a variety of kefir with a distinct sour taste. It is usually mixed with fresh herbs in a pesto-like purée called delal. Yogurt is a side dish to all Iranian meals. Strained yogurt is used as dips and various appetizers with multitudes of ingredients: cucumbers, onions, shallots, fresh herbs (dill, spearmint, parsley, cilantro), spinach, walnuts, zereshk, garlic, etc. The most popular appetizers are spinach or eggplant borani, "Mâst-o-Khiâr" with cucumber, spring onions and herbs, or "Mâst-Musir" with wild shallots.

In Egypt, strained and unstrained yogurt is called "zabadi" ("laban" meaning "milk" in Egyptian Arabic). It is eaten with savoury accompaniments such as olives and oil, and also with a sweetener such as honey, as a snack or breakfast food. Areesh cheese (or Arish) is a type of cheese that originated in Egypt. Shanklish, a fermented cheese, is made from areesh cheese. Arish cheese is made from yogurt heated slowly until it curdles and separates, then placed in cheesecloth to drain. It is similar in taste to Ricotta. The protein content of Areesh cheese is 17.6%.

North America

Strained yogurt (often marketed as "Greek yogurt") has become popular in the United States and Canada, where it is often used as a lower-calorie substitute for sour cream or crème fraîche. Celebrity chef Graham Kerr became an early adopter of strained yogurt as an ingredient, frequently featuring it (and demonstrating how to strain plain yogurt through a coffee filter) on his eponymous 1990 cooking show, as frequently as he had featured clarified butter on *The Galloping Gourmet* in the late 1960s. In 2015, food market research firm Packaged Facts reported that Greek yogurt has a 50 percent share of the yogurt market in the United States.

"Greek yogurt" brands in North America include Chobani, Dannon Oikos, FAGE, Olympus, Stonyfield organic Oikos, Yoplait, Cabot Creamery and Voskos. FAGE began importing its Greek products in 1998 and opened a domestic production plant in Johnstown, New York, in 2008. Chobani, based in New Berlin, New York, began marketing its Greek-style yogurt in 2007. The Voskos brand entered the US market in 2009 with imported Greek yogurt products at 10%, 2%, and 0% milkfat. Stonyfield Farms, owned by Groupe Danone, introduced Oikos Organic Greek Yogurt in 2007; Danone began marketing a non-organic Dannon Oikos Greek Yogurt in 2011 and also produced a now discontinued blended Greek-style yogurt under the Activia Selects brand; Dannon Light & Fit Greek nonfat yogurt was introduced in 2012 and boasts being the lightest Greek yogurt with fruit, and Activia Greek yogurt was re-introduced in 2013. General Mills introduced a Greek-style yogurt under the Yoplait brand name in early 2010, which was discontinued and replaced by Yoplait Greek 100 in August 2012. Activia Greek yogurt was re-introduced in 2013, and in July 2012 took over US distribution and sales of Canadian Liberté's Greek brands. In Canada, Yoplait was launched in January 2013, and is packaged with toppings.

Mexico

Strained yogurt is called *jocoque seco* in Mexico. It was popularised by local producers of Lebanese origin and is widely popular in the country.

Northern Europe

Strained yogurt, in full-, low-, and no-fat versions, has become popular in northern European cookery as a low-calorie alternative to cream in recipes. It is typically marketed as "Greek" or "Turkish" yogurt.

In Denmark, a type of strained yogurt named ymer is available. In contrast to the Greek and Turkish variety, only a minor amount of whey is drained off in the production process. Ymer is traditionally consumed with the addition of ymerdrys *(lit. Danish:* ymer sprinkle*)*, a mixture of bread crumbs made from rye bread (*rugbrød*) and brown sugar. Like other types of soured dairy products, ymer is often consumed at breakfast. Strained yogurt topped with muesli and maple syrup is often served at brunch in cafés in Denmark.

In the Netherlands, strained yogurt is known as *hangop*, literally meaning 'hang up'. It is a traditional dessert. *Hangop* may also be made using buttermilk.

Turkey

In Turkey, strained yogurt is known as *süzme yoğurt* ("strained yogurt") or *kese yoğurdu* ("bag yogurt"). Water is sometimes added to it in the preparation of cacık, when this is not eaten as a meze but consumed as a beverage. Strained yogurt is used in Turkish mezzes and dips such as haydari.

Unstirred Turkish *Süzme Yoğurt* (strained yogurt), with a 10% fat content

In Turkish markets, labne is also a popular dairy product but it is different from strained yogurt; it is yogurt-based creamy cheese without salt, and is used like mascarpone.

United Kingdom

In the United Kingdom strained yogurt can only be marketed as "Greek" if made in Greece. Strained cows'-milk yogurt not made in Greece is typically sold as "Greek Style" or "Greek Recipe" for marketing reasons, typically at lower prices than yogurt made in Greece. Among "Greek Style" yogurts there is no distinction between those thickened by straining and those thickened through additives.

In September 2012 Chobani UK Ltd. began to sell yogurt made in the United States as "Greek Yogurt". FAGE, a company that manufactures yogurt in Greece and sells it in the UK, filed a passing-off claim against Chobani in the UK High Court, claiming that UK consumers understood "Greek" to refer to the country of origin (similar to "Belgian Beer"); Chobani's position was that consumers understood "Greek" to refer to a preparation (similar to "French Toast"). Both companies relied on surveys to prove their point; FAGE also relied on the previous industry practice of UK yogurt makers to not label their yogurt as "Greek Yogurt". Ultimately Mr Justice Briggs found in favor of FAGE and granted an injunction preventing Chobani from using the name "Greek Yogurt". In February 2014 this decision was upheld on appeal. Chobani later announced that it was reentering the UK market using a "strained yogurt" label but has not yet done so.

Production Issues

The characteristic thick texture and high protein content are achieved through either or both of two processing steps. The milk may be concentrated by ultrafiltration to remove a portion of the water before addition of yogurt cultures. Alternatively, after culturing, the yogurt may be centrifuged or membrane-filtered to remove whey, in a process analogous to the traditional straining step. Brands described as "strained" yogurt, including

Activia Greek, Chobani, Dannon Light & Fit Greek, Dannon Oikos, FAGE, Stonyfield Organic Oikos, Trader Joe's, and Yoplait have undergone the second process. Process details are highly guarded trade secrets. Other brands of Greek-style yogurt, including Yoplait and some store brands, are made by adding milk protein concentrate and thickeners to standard yogurt to boost the protein content and modify the texture.

The liquid resulting from straining yogurt is called "acid whey" and is composed of water, yogurt cultures, protein, a slight amount of lactose, and lactic acid. It is difficult to dispose of. Farmers have used the whey to mix with animal feed and fertilizer. Using anaerobic digesters, it can be a source of methane that can be used to produce electricity.

Doogh

Bottle of carbonated *tan* sold in Yerevan, Armenia

Doogh, Ayran or Tan is a savory yogurt-based beverage. Doogh is the same drink known in Turkey as ayran. It is popular in Iran, Turkey, Azerbaijan, Armenia, Kazakhstan, Kyrgystan, North Caucasus, the Balkans, Afghanistan (by the Kirghiz) and Lebanon. It is made by mixing yoghurt and chilled or iced water and has been variously described as "diluted yogurt". It is sometimes carbonated and seasoned with mint.

History

According to Shirin Simmons, *doogh* has long been a popular drink and was consumed in ancient Persia (modern-day Iran). Described by an 1886 source as a cold drink of curdled milk and water seasoned with mint, its name derives from the Persian word for milking, *dooshidan*.

According to Nevin Halici, *ayran* is a traditional Turkish drink and was consumed by nomadic Turks prior to 1000 CE. According to Celalettin Koçak and Yahya Kemal Avsar (Professor of Food Engineering at Mustafa Kemal University), *ayran* was first developed thousands of years ago by the Göktürks, who would dilute bitter yogurt with water in an attempt to improve its flavor.

A c. 1000 CE Turkish dictionary, Dīwān ul-Lughat al-Turk, defines ayran as a "drink made out of milk."

Turkish National Drink Status

Recep Tayyip Erdoğan, a Turkish politician who has held the posts of President and Prime Minister, has promoted ayran as a national drink. Speaking at a 2013 WHO Global Alcohol Policy Conference held in İstanbul, Erdoğan contrasted ayran with alcohol, which he suggested was a recent introduction to Turkey. Stating that in the early years of the modern Turkish republic (c. 1920-1950), alcoholic beverages were "part of the radical top-down modernization program embarked upon by the elites," Erdoğan expressed regret that alcohol was widely promoted during this period even in school textbooks.

Still, sales of ayran in Turkey may lag behind other non-alcoholic beverages. According to a 2015 joint statement from the Soft Drink Producers Association, the Sparkling Water Producers Association, and the Milk Producers and Exporters Union of Turkey, ayran consumption during the holy month of Ramadan has declined every year for the years 2010-15.

In 2015, Turkey's Customs and Trade Ministry imposed a 220,000 Turkish Lira fine (approx. $70,000) to state-owned Çaykur manufacturers for "insulting ayran" in one of their advertisement for iced tea, in which the protagonist raps that ayran makes him sleepy, and halted advertisements of Çaykur's competing, ice-tea product.

Variations

Dhallë is a traditional cold drink in Albania made of yogurt and water

Salt (and sometimes pepper) is added, and dried mint or pennyroyal can be mixed in as well. One variation includes diced cucumbers to provide a crunchy texture to the beverage. Some varieties of *doogh* have carbonation.

Curd

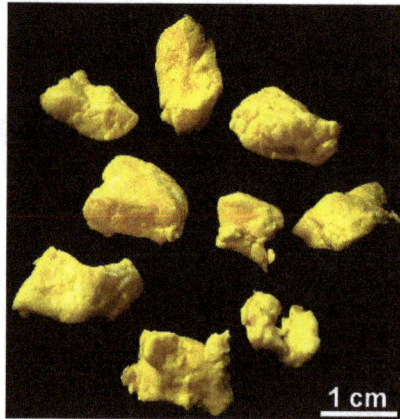

Cheese curds

Curds are a dairy product obtained by coagulating milk in a process called curdling. The coagulation can be caused by adding rennet or any edible acidic substance such as lemon juice or vinegar, and then allowing it to sit. The increased acidity causes the milk proteins (casein) to tangle into solid masses, or *curds*. Milk that has been left to sour (raw milk alone or pasteurized milk with added lactic acid bacteria) will also naturally produce curds, and sour milk cheeses are produced this way. Producing cheese curds is one of the first steps in cheesemaking; the curds are pressed and drained to varying amounts for different styles of cheese and different secondary agents (molds for blue cheeses, etc.) are introduced before the desired aging finishes the cheese. The remaining liquid, which contains only whey proteins, is the whey. In cow's milk, 80% of the proteins are caseins.

Uses

Lithuanian curd

Curd products vary by region and include cottage cheese, curd cheese (both curdled by bacteria and sometimes also rennet), farmer cheese, pot cheese, queso blanco, and Indian *paneer* (milk curdled with lime juice). The word can also refer to a non-dairy substance of similar appearance or consistency, though in these cases a modifier or the word *curdled* is generally used.

In England, curds produced from the use of rennet are referred to as junket, with true curds and whey only occurring from the natural separation of milk due to its environment (temperature, acidity).

Cheese curds, drained of the whey and served without further processing or aging, are popular in some French-speaking regions of Canada, such as Quebec, parts of Ontario, and Atlantic Canada. In Quebec, Eastern Ontario and the Eastern provinces such as New Brunswick, cheese curds are popularly served with french fries and gravy as *poutine*. In some parts of the Midwestern U.S., especially in Wisconsin, they are breaded and fried, or are eaten straight.

In Turkey, curds are called *keş* and are very commonly used as an aphrodisiac and for breakfast served on fried bread and are also eaten with macaroni in the provinces of Bolu and Zonguldak.

In Mexico, the chongos zamoranos is a dessert prepared with milk curdled with sugar and cinnamon.

Albanian *gjiza* is made by boiling whey for about 15 minutes and adding vinegar or lemon. The derivative is drained 3 to 4 times with a napkin or piece of cloth and salted to taste. Gjiza can be served immediately or refrigerated for a couple of days.

Formation

Old Dutch Cheese press

Lactobacillus is a genus of bacteria which can convert sugars into lactic acid by means of fermentation. Milk contains a sugar called lactose, a disaccharide (compound sugar) made by the glycosidic bonding between glucose and galactose (monosaccharides).

When pasteurized milk is heated to a temperature of 30-40 °C, or even at room temperature or refrigerator temperature, and a small amount of old curd or whey added to it, the lactobacillus in that curd or whey sample starts to grow. These convert the lactose into lactic acid, which imparts the sour taste to curd. Raw milk naturally contains lactobacillus.

Curds in Song and Poetry

Curds are mentioned in the Middle Irish parodic tale *Aislinge Meic Con Glinne* (The Vision of Mac Conglinne), the relevant portion of which reads:

> Stately, pleasantly it sat,
> A compact house and strong.
> Then I went in:
> The door of it was dry meat,
> The threshold was bare bread,
> cheese-curds the sides.
>
> Smooth pillars of old cheese,
> And sappy bacon props
> Alternate ranged;
> Fine beams of mellow cream,
> White rafters - real curds,
> Kept up the house.

Curds also appear in the nursery rhyme *Little Miss Muffet*.

Buffalo Curd

A pot of buffalo curd with treacle in Sri Lanka

Buffalo curd is a traditional type of yogurt prepared from buffalo milk. It is popular throughout south Asian countries such as Bangladesh, India, Nepal, Pakistan, Sri Lanka, etc. Buffalo milk is traditionally better than cow milk due to its higher fat content making a thicker yogurt mass. Mostly clay pots are used as packaging material for Buffalo curd. The naming *curd* is traditionally used in the Indian subcontinent to refer to yogurt, while another word, *paneer*, is used to denote curd in the British English word sense.

A cup of curd ready for the dessert

Buffalo curd is obtained by bacterial fermentation of buffalo milk. In this process lactose in buffalo milk is converted into lactic acid using several micro-organisms. The species involved in the fermentation include *Lactococcus lactis, Streptococcus diacetylactis, Streptococcus cremoris, Lactobacillus delbrueckii subsp. bulgaricus* and *Streptococcus thermophilus.*

Buffalo curd has a higher nutritional value of protein, fat, lactose, minerals and vitamins. It should have 7.5% of milk fat, 8.5% of milk solids and 4.5% of Milk acid (lactic acid). Quality of the curd totally depends on the starter culture. Fermentation also develops the characteristic flavor and colour of the product.

Buffalo curd can be made in both traditional and industrial forms. Traditionally buffalo milk is filtered and boiled, the scum is removed and it is cooled to room temperature. A few spoonfuls of a previous batch of curd are added and it is then mixed well and poured into clay pots. These are sealed by wrapping a piece of paper over the pot and allowing it to stand for 12 hours.

Bangladesh (Bhola Island)

Bhola Island is known for its Buffalo curd (*doi*) which is unique in Bangladesh, as most curds are made from cattle milk. Raw buffalo milk is used to make the curd. The process that has been used has remained unchanged. It is made in traditional clay pots and the process takes 18 hours. The common method of making curd in Bangladesh is using a sample of the old curd. Buffalo milk which has a higher fat content than cow milk congeals on its own. To help the process only raw milk is used. It is popular in the Island and is served in special occasions such as weddings, Pooja and other religious and cultural occasions.

References

- Peters, Pam (2004). The Cambridge Guide to English Usage. Cambridge: Cambridge University Press, pp. 587–588, ISBN 052162181X.

- Don Tribby. Yoghurt. Chapter 8 in The Sensory Evaluation of Dairy Products. Eds. Stephanie Clark, et al. Springer Science & Business Media, 2009 ISBN 9780387774084 Page 191

- Batmanglij, Najmieh (2007). A Taste of Persia: An Introduction to Persian Cooking. I.B.Tauris. p. 170. ISBN 978-1-84511-437-4.

- Farnworth, Edward R. (2008). Handbook of fermented functional foods. Taylor and Francis. p. 114. ISBN 978-1-4200-5326-5.

- Coyle, L. Patrick (1982). The World Encyclopedia of Food. Facts On File Inc. p. 763. ISBN 978-0-87196-417-5. Retrieved 11 August 2009.

- Hui, ed. Ramesh C. Chandan, associate editors, Charles H. White, Arun Kilara, Y. H. (2006). Manufacturing yogurt and fermented milks (1. ed.). Ames (Iowa): Blackwell. p. 364. ISBN 9780813823041.

- Kelley, Laura (2009). The Silk Road Gourmet: Western and Southern Asia. New York: iUniverse. p. 191. ISBN 9781440143052.

- Walker, Harlan, ed. (2000) Milk-- Beyond the Dairy: Proceedings of the Oxford Symposium on Food and Cookery 1999 Totnes, Devon, Eng. : Prospect Books. page 276. ISBN 9781903018064.

- Sarina Jacobson,Danya Weiner. Yogurt: More Than 70 Delicious & Healthy Recipes" Sterling Publishing Company, Inc., 2008. ISBN 1402747594 p 6

- Yildiz Fatih (2010). Development and Manufacture of Yogurt and Other Functional Dairy Products. CRC Press. p. 10. ISBN 9781420082081.

Cheese and Cheese Making Processes

Cheese is a food item that is derived from milk and is produced in different flavors. There are different types of cheese; almost 500 different varieties of cheese are produced. The process used in making cheese is known as cheese ripening. This section will provide an integrated understanding of cheese and cheese making processes.

Cheese

Cheese is a food derived from milk that is produced in a wide range of flavors, textures, and forms by coagulation of the milk protein casein. It comprises proteins and fat from milk, usually the milk of cows, buffalo, goats, or sheep. During production, the milk is usually acidified, and adding the enzyme rennet causes coagulation. The solids are separated and pressed into final form. Some cheeses have molds on the rind or throughout. Most cheeses melt at cooking temperature.

A platter with cheese and garnishes

Hundreds of types of cheese from various countries are produced. Their styles, textures and flavors depend on the origin of the milk (including the animal's diet), whether they have been pasteurized, the butterfat content, the bacteria and mold, the processing, and aging. Herbs, spices, or wood smoke may be used as flavoring agents. The yellow to red color of many cheeses, such as Red Leicester, is produced by adding annatto. Other ingredients may be added to some cheeses, such as black pepper, garlic, chives or cranberries.

For a few cheeses, the milk is curdled by adding acids such as vinegar or lemon juice. Most cheeses are acidified to a lesser degree by bacteria, which turn milk sugars into lactic acid, then the addition of rennet completes the curdling. Vegetarian alternatives to rennet are available; most are produced by fermentation of the fungus *Mucor miehei*, but others have been extracted from various species of the *Cynara* thistle family. Cheesemakers near a dairy region may benefit from fresher, lower-priced milk, and lower shipping costs.

A variety of cheeses

Coulommiers cheese

Cheese is valued for its portability, long life, and high content of fat, protein, calcium, and phosphorus. Cheese is more compact and has a longer shelf life than milk, although how long a cheese will keep depends on the type of cheese; labels on packets of cheese often claim that a cheese should be consumed within three to five days of opening. Generally speaking, hard cheeses, such as parmesan last longer than soft cheeses, such as Brie or goat's milk cheese. The long storage life of some cheeses, especially when encased in a protective rind, allows selling when markets are favorable.

There is some debate as to the best way to store cheese, but some experts say that wrapping it in cheese paper provides optimal results. Cheese paper is coated in a porous plastic on the inside, and the outside has a layer of wax. This specific combination of plastic on the inside and wax on the outside protects the cheese by allowing condensation on the cheese to be wicked away while preventing moisture from within the cheese escaping.

A specialist seller of cheese is sometimes known as a *cheesemonger*. Becoming an expert in this field requires some formal education and years of tasting and hands-on experience, much like becoming an expert in wine or cuisine. The cheesemonger is responsible for all aspects of the cheese inventory: selecting the cheese menu, purchasing, receiving, storage, and ripening.

Etymology

Cheese on market stand in Basel, Switzerland

The word *cheese* comes from Latin *caseus*, from which the modern word casein is also derived. The earliest source is from the proto-Indo-European root **kwat-*, which means "to ferment, become sour".

More recently, *cheese* comes from *chese* (in Middle English) and *cīese* or *cēse* (in Old English). Similar words are shared by other West Germanic languages—West Frisian *tsiis*, Dutch *kaas*, German *Käse*, Old High German *chāsi*—all from the reconstructed West-Germanic form **kāsī*, which in turn is an early borrowing from Latin.

When the Romans began to make hard cheeses for their legionaries' supplies, a new word started to be used: *formaticum*, from *caseus formatus*, or "molded cheese" (as in "formed", not "moldy"). It is from this word that the French *fromage*, proper Italian *formaggio*, Catalan *formatge*, Breton *fourmaj*, and Provençal *furmo* are derived. Of the Romance languages, Spanish, Portuguese, Romanian, Tuscan and Southern Italian dialects use words derived from *caseus*. The word *cheese* itself is occasionally employed in a sense that means "molded" or "formed". *Head cheese* uses the word in this sense.

Origins

A piece of soft curd cheese, oven-baked to increase longevity

Cheese is an ancient food whose origins predate recorded history. There is no conclusive evidence indicating where cheesemaking originated, either in Europe, Central Asia or the Middle East, but the practice had spread within Europe prior to Roman times and, according to Pliny the Elder, had become a sophisticated enterprise by the time the Roman Empire came into being.

The earliest evidence of cheese-making in the archaeological record dates back to 5,500 BCE, in what is now Kujawy, Poland, where strainers with milk fats molecules have been found. Earliest proposed dates for the origin of cheesemaking range from around 8000 BCE, when sheep were first domesticated. Since animal skins and inflated internal organs have, since ancient times, provided storage vessels for a range of foodstuffs, it is probable that the process of cheese making was discovered accidentally by storing milk in a container made from the stomach of an animal, resulting in the milk being turned to curd and whey by the rennet from the stomach. There is a legend – with variations – about the discovery of cheese by an Arab trader who used this method of storing milk.

Cheesemaking may have begun independently of this by the pressing and salting of curdled milk to preserve it. Observation that the effect of making cheese in an animal stomach gave more solid and better-textured curds may have led to the deliberate addition of rennet.

Early archeological evidence of Egyptian cheese has been found in Egyptian tomb murals, dating to about 2000 BCE. The earliest cheeses were likely to have been quite sour and salty, similar in texture to rustic cottage cheese or feta, a crumbly, flavorful Greek cheese.

Cheese produced in Europe, where climates are cooler than the Middle East, required less salt for preservation. With less salt and acidity, the cheese became a suitable environment for useful microbes and molds, giving aged cheeses their respective flavors.

The earliest ever discovered preserved cheese was found in the Taklamakan Desert in Xinjiang, China, and it dates back as early as 1615 BCE.

Ancient Greece and Rome

Cheese in a market in Italy

Ancient Greek mythology credited Aristaeus with the discovery of cheese. Homer's *Odyssey* (8th century BCE) describes the Cyclops making and storing sheep's and goats' milk cheese (translation by Samuel Butler):

We soon reached his cave, but he was out shepherding, so we went inside and took stock of all that we could see. His cheese-racks were loaded with cheeses, and he had more lambs and kids than his pens could hold...

When he had so done he sat down and milked his ewes and goats, all in due course, and then let each of them have her own young. He curdled half the milk and set it aside in wicker strainers.

Cheese, Tacuinum sanitatis Casanatensis (14th century)

By Roman times, cheese was an everyday food and cheesemaking a mature art. Columella's *De Re Rustica* (circa 65 CE) details a cheesemaking process involving rennet coagulation, pressing of the curd, salting, and aging. Pliny's *Natural History* (77 CE) devotes a chapter (XI, 97) to describing the diversity of cheeses enjoyed by Romans of the early Empire. He stated that the best cheeses came from the villages near Nîmes, but did not keep long and had to be eaten fresh. Cheeses of the Alps and Apennines were as remarkable for their variety then as now. A Ligurian cheese was noted for being made mostly from sheep's milk, and some cheeses produced nearby were stated to weigh as much as a thousand pounds each. Goats' milk cheese was a recent taste in Rome, improved over the "medicinal taste" of Gaul's similar cheeses by smoking. Of cheeses from overseas, Pliny preferred those of Bithynia in Asia Minor.

Post-roman Europe

As Romanized populations encountered unfamiliar newly settled neighbors, bringing their own cheese-making traditions, their own flocks and their own unrelated words for *cheese*, cheeses in Europe diversified further, with various locales developing their own distinctive traditions and products. As long-distance trade collapsed, only travelers would encounter unfamiliar cheeses: Charlemagne's first encounter with a white cheese that had an edible rind forms one of the constructed anecdotes of Notker's *Life* of the Emperor.

The British Cheese Board claims that Britain has approximately 700 distinct local cheeses; France and Italy have perhaps 400 each. (A French proverb holds there is a different French cheese for every day of the year, and Charles de Gaulle once asked "how can you govern a country in which there are 246 kinds of cheese?") Still, the advancement of the cheese art in Europe was slow during the centuries after Rome's fall. Many cheeses today were first recorded in the late Middle Ages or after—cheeses like Cheddar around 1500, Parmesan in 1597, Gouda in 1697, and Camembert in 1791.

In 1546 *The Proverbs of John Heywood* claimed "the moon is made of a greene cheese." (*Greene* may refer here not to the color, as many now think, but to being new or unaged.) Variations on this sentiment were long repeated and NASA exploited this myth for an April Fools' Day spoof announcement in 2006.

Modern Era

Until its modern spread along with European culture, cheese was nearly unheard of in east Asian cultures, in the pre-Columbian Americas, and only had limited use in sub-Mediterranean Africa, mainly being widespread and popular only in Europe, the Middle East, the Indian subcontinent, and areas influenced by those cultures. But with the spread, first of European imperialism, and later of Euro-American culture and food, cheese has gradually become known and increasingly popular worldwide.

The first factory for the industrial production of cheese opened in Switzerland in 1815, but large-scale production first found real success in the United States. Credit usually goes to Jesse Williams, a dairy farmer from Rome, New York, who in 1851 started making cheese in an assembly-line fashion using the milk from neighboring farms. Within decades, hundreds of such dairy associations existed.

The 1860s saw the beginnings of mass-produced rennet, and by the turn of the century scientists were producing pure microbial cultures. Before then, bacteria in cheesemaking had come from the environment or from recycling an earlier batch's whey; the pure cultures meant a more standardized cheese could be produced.

Factory-made cheese overtook traditional cheesemaking in the World War II era, and factories have been the source of most cheese in America and Europe ever since.

Cheese production – 2013	
Country	**Production (millions of tonnes)**
United States	5.4
Germany	2.2
France	1.9
Italy	1.2
Netherlands	0.8
World	**21.3**
Source: FAOSTAT of the United Nations	

Production

In 2013, world production of cheese was 21.3 million tonnes, with the United States accounting for 25% (5.4 million tonnes) of the world total followed by Germany, France and Italy (table).

During 2015, Germany, France, Netherlands and Italy exported 10-14% of their produced cheese. The United States, the biggest world producer of cheese, is a marginal exporter (5.3% of total), as most of its production is for the domestic market.

Consumption

France, Iceland, Finland, Denmark and Germany were the highest consumers of cheese in 2014, averaging 25 kg (55 lb) per person.

Processing

Curdling

A required step in cheesemaking is separating the milk into solid curds and liquid whey. Usually this is done by acidifying (souring) the milk and adding rennet. The acidification can be accomplished directly by the addition of an acid, such as vinegar, in a few cases (paneer, queso fresco). More commonly starter bacteria are employed instead which convert milk sugars into lactic acid. The same bacteria (and the enzymes they produce) also play a large role in the eventual flavor of aged cheeses. Most cheeses are made with starter bacteria from the *Lactococcus*, *Lactobacillus*, or *Streptococcus* families. Swiss starter cultures also include *Propionibacter shermani*, which produces carbon dioxide gas bubbles during aging, giving Swiss cheese or Emmental its holes (called "eyes").

During industrial production of Emmental cheese, the as-yet-undrained curd is broken by rotating mixers.

Some fresh cheeses are curdled only by acidity, but most cheeses also use rennet. Rennet sets the cheese into a strong and rubbery gel compared to the fragile curds produced by acidic coagulation alone. It also allows curdling at a lower acidity—important because flavor-making bacteria are inhibited in high-acidity environments. In general, softer, smaller, fresher cheeses are curdled with a greater proportion of acid to rennet than harder, larger, longer-aged varieties.

While rennet was traditionally produced via extraction from the inner mucosa of the fourth stomach chamber of slaughtered young, unweaned calves, most rennet used today in cheesemaking is produced recombinantly. The majority of the applied chymosin is retained in the whey and, at most, may be present in cheese in trace quantities. In ripe cheese, the type and provenance of chymosin used in production cannot be determined.

Curd Processing

At this point, the cheese has set into a very moist gel. Some soft cheeses are now essentially complete: they are drained, salted, and packaged. For most of the rest, the curd is cut into small cubes. This allows water to drain from the individual pieces of curd.

Some hard cheeses are then heated to temperatures in the range of 35–55 °C (95–131 °F). This forces more whey from the cut curd. It also changes the taste of the finished cheese, affecting both the bacterial culture and the milk chemistry. Cheeses that are heated to the higher temperatures are usually made with thermophilic starter bacteria that survive this step—either *Lactobacilli* or *Streptococci*.

Salt has roles in cheese besides adding a salty flavor. It preserves cheese from spoiling, draws moisture from the curd, and firms cheese's texture in an interaction with its proteins. Some cheeses are salted from the outside with dry salt or brine washes. Most cheeses have the salt mixed directly into the curds.

Cheese factory in the Netherlands

Other techniques influence a cheese's texture and flavor. Some examples are :

- Stretching: (Mozzarella, Provolone) The curd is stretched and kneaded in hot water, developing a stringy, fibrous body.

- Cheddaring: (Cheddar, other English cheeses) The cut curd is repeatedly piled up, pushing more moisture away. The curd is also mixed (or *milled*) for a long time, taking the sharp edges off the cut curd pieces and influencing the final product's texture.

- Washing: (Edam, Gouda, Colby) The curd is washed in warm water, lowering its acidity and making for a milder-tasting cheese.

Most cheeses achieve their final shape when the curds are pressed into a mold or form. The harder the cheese, the more pressure is applied. The pressure drives out moisture—the molds are designed to allow water to escape—and unifies the curds into a single solid body.

Parmigiano-Reggiano in a modern factory

Ripening

A newborn cheese is usually salty yet bland in flavor and, for harder varieties, rubbery in texture. These qualities are sometimes enjoyed—cheese curds are eaten on their own—but normally cheeses are left to rest under controlled conditions. This aging period (also called ripening, or, from the French, *affinage*) lasts from a few days to several years. As a cheese ages, microbes and enzymes transform texture and intensify flavor. This transformation is largely a result of the breakdown of casein proteins and milkfat into a complex mix of amino acids, amines, and fatty acids.

Some cheeses have additional bacteria or molds intentionally introduced before or during aging. In traditional cheesemaking, these microbes might be already present in the aging room; they are simply allowed to settle and grow on the stored cheeses. More often today, prepared cultures are used, giving more consistent results and putting fewer constraints on the environment where the cheese ages. These cheeses include soft ripened cheeses such as Brie and Camembert, blue cheeses such as Roquefort, Stilton, Gorgonzola, and rind-washed cheeses such as Limburger.

Types

Feta from Greece

Local cheese at an open-air market in Peru.

There are many types of cheese, with around 500 different varieties recognized by the International Dairy Federation, more than 400 identified by Walter and Hargrove, more than 500 by Burkhalter, and more than 1,000 by Sandine and Elliker. The varieties may be grouped or classified into types according to criteria such as length of ageing, texture, methods of making, fat content, animal milk, country or region of origin, etc.—with these criteria either being used singly or in combination, but with no single method being universally used. The method most commonly and traditionally used is based on moisture content, which is then further discriminated by fat content and curing or ripening methods. Some attempts have been made to rationalise the classification of cheese—a scheme was proposed by Pieter Walstra which uses the primary and secondary starter combined with moisture content, and Walter and Hargrove suggested classifying by production methods which produces 18 types, which are then further grouped by moisture content.

Moisture Content (Soft to Hard)

Categorizing cheeses by firmness is a common but inexact practice. The lines between "soft", "semi-soft", "semi-hard", and "hard" are arbitrary, and many types of cheese are made in softer or firmer variations. The main factor that controls cheese hardness is moisture content, which depends largely on the pressure with which it is packed into molds, and on aging time.

Fresh, Whey and Stretched Curd Cheeses

The main factor in the categorization of these cheeses is their age. Fresh cheeses without additional preservatives can spoil in a matter of days.

Emmental

Some cheeses are categorized by the source of the milk used to produce them or by the added fat content of the milk from which they are produced. While most of the world's commercially available cheese is made from cows' milk, many parts of the world also produce cheese from goats and sheep. Double cream cheeses are soft cheeses of cows' milk enriched with cream so that their fat content is 60% or, in the case of triple creams, 75%. The use of the terms "double" or "triple" is not meant to give a quantitative reference to the change in fat content, since the fat content of whole cows' milk is 3%-4%.

Soft-ripened and Blue-vein

There are at least three main categories of cheese in which the presence of mold is a significant feature: soft ripened cheeses, washed rind cheeses and blue cheeses.

Processed Cheeses

Processed cheese is made from traditional cheese and emulsifying salts, often with the addition of milk, more salt, preservatives, and food coloring. It is inexpensive, consistent, and melts smoothly. It is sold packaged and either pre-sliced or unsliced, in a number of varieties. It is also available in aerosol cans in some countries.

Cooking and Eating

Zigerbrüt, cheese grated onto bread through a mill, from the Canton of Glarus in Switzerland.

Saganaki, lit on fire, served in Chicago.

At refrigerator temperatures, the fat in a piece of cheese is as hard as unsoftened butter, and its protein structure is stiff as well. Flavor and odor compounds are less easily liberated when cold. For improvements in flavor and texture, it is widely advised that cheeses be allowed to warm up to room temperature before eating. If the cheese is further warmed, to 26–32 °C (79–90 °F), the fats will begin to "sweat out" as they go beyond soft to fully liquid.

Above room temperatures, most hard cheeses melt. Rennet-curdled cheeses have a gel-like protein matrix that is broken down by heat. When enough protein bonds are broken, the cheese itself turns from a solid to a viscous liquid. Soft, high-moisture cheeses will melt at around 55 °C (131 °F), while hard, low-moisture cheeses such as Parmesan remain solid until they reach about 82 °C (180 °F). Acid-set cheeses, including halloumi, paneer, some whey cheeses and many varieties of fresh goat cheese, have a protein structure that remains intact at high temperatures. When cooked, these cheeses just get firmer as water evaporates.

Some cheeses, like raclette, melt smoothly; many tend to become stringy or suffer from a separation of their fats. Many of these can be coaxed into melting smoothly in the presence of acids or starch. Fondue, with wine providing the acidity, is a good example of a smoothly melted cheese dish. Elastic stringiness is a quality that is sometimes enjoyed, in dishes including pizza and Welsh rarebit. Even a melted cheese eventually turns solid again, after enough moisture is cooked off. The saying "you can't melt cheese twice" (meaning "some things can only be done once") refers to the fact that oils leach out during the first melting and are gone, leaving the non-meltable solids behind.

As its temperature continues to rise, cheese will brown and eventually burn. Browned, partially burned cheese has a particular distinct flavor of its own and is frequently used in cooking (e.g., sprinkling atop items before baking them).

Nutrition and Health

Mozzarella

The nutritional value of cheese varies widely. Cottage cheese may consist of 4% fat and 11% protein while some whey cheeses are 15% fat and 11% protein, and triple-crème cheeses are 36% fat and 7% protein. In general, cheese is a rich source (20% or more of the Daily Value, DV) of calcium, protein, phosphorus, sodium and saturated fat. A 28-gram (0.99 oz) (one ounce) serving of cheddar cheese contains about 7 grams (0.25 oz) of protein and 202 milligrams of calcium. Nutritionally, cheese is essentially concentrated milk: it takes about 200 grams (7.1 oz) of milk to provide that much protein, and 150 grams (5.3 oz) to equal the calcium.

MacroNutrients (grams) of common cheeses per 100gm				
Cheese	**Water**	**Protein**	**Fat**	**Carbs**
Swiss	37.1	26.9	27.8	5.4
Feta	55.2	14.2	21.3	4.1
Cheddar	36.8	24.9	33.1	1.3
Mozarella	50	22.2	22.4	2.2
Cottage	80	11.1	4.3	3.4

Vitamin contents in %DV of common cheeses per 100gm													
Cheese	**A**	**B1**	**B2**	**B3**	**B5**	**B6**	**B9**	**B12**	**Ch.**	**C**	**D**	**E**	**K**
Swiss	17	4	17	0	4	4	1	**56**	2.8	0	11	2	3
Feta	8	10	50	5	10	21	8	28	2.2	0	0	1	2
Cheddar	20	2	22	0	4	4	5	14	3	0	3	1	3
Mozzarella	14	2	17	1	1	2	2	38	2.8	0	0	1	3
Cottage	3	2	10	0	6	2	3	7	3.3	0	0	0	0

Mineral contents in %DV of common cheeses per 100 grams										
Cheese	**Ca**	**Fe**	**Mg**	**P**	**K**	**Na**	**Zn**	**Cu**	**Mn**	**Se**
Swiss	79	10	1	57	2	8	29	2	0	26
Feta	49	4	5	34	2	46	19	2	1	21
Cheddar	72	4	7	51	3	26	21	2	1	20
Mozzarella	51	2	5	35	2	26	19	1	1	24
Cottage	8	0	2	16	3	15	3	1	0	14

Ch. = Choline; Ca = Calcium; Fe = Iron; Mg = Magnesium; P = Phosphorus; K = Potassium; Na = Sodium; Zn = Zinc; Cu = Copper; Mn = Manganese; Se = Selenium;

Note : All nutrient values including protein are in %DV per 100 grams of the food item except for Macro-nutrients. Source : Nutritiondata.self.com

Neonatal Infection and Death

Cheese has the potential for promoting the growth of *Listeria* bacteria. *Listeria monocytogenes* can also cause serious infection in an infant and pregnant woman and can be transmitted to her infant in utero or after birth. The infection has the potential of seriously harming or even causing the death of a preterm infant, an infant of low or very low birth weight, or an infant with an immune system deficiency or a congenital defect of the immune system. The presence of this pathogen can sometimes be determined by the symptoms that appear as a gastrointestinal illness in the mother. The mother can also acquire infection from ingesting food that contains other animal products such as, unpasteurized milk, delicatessen meats, and hot dogs.

Heart Disease

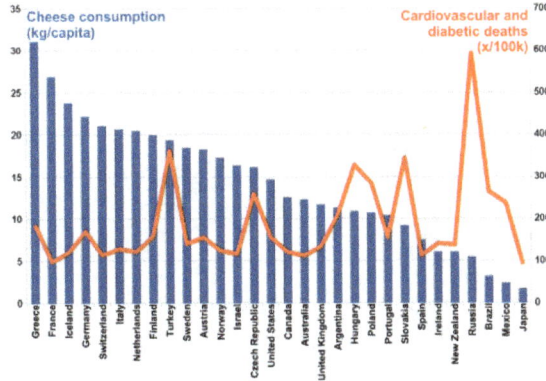

Average cheese consumption and rates of mortality due to cardiovascular disease or diabetes

A review of the medical literature published in 2012 noted that: "Cheese consumption is the leading contributor of SF (saturated fat) in the U.S. diet, and therefore would be predicted to increase LDL-C (LDL cholesterol) and consequently increase the risk of CVD (cardiovascular disease)." It found that: "Based on results from numerous prospective observational studies and meta-analyses, most, but not all, have shown no association and in some cases an inverse relationship between the intake of milk fat containing dairy products and the risk of CVD, CHD (coronary heart disease), and stroke. A limited number of prospective cohort studies found no significant association between the intake of total full-fat dairy products and the risk of CHD or stroke....Most clinical studies showed that full-fat natural cheese, a highly fermented product, significantly lowers LDL-C compared with butter intake of equal total fat and saturated fat content."

Pasteurization

A number of food safety agencies around the world have warned of the risks of raw-milk cheeses. The U.S. Food and Drug Administration states that soft raw-milk cheeses can cause "serious infectious diseases including listeriosis, brucellosis, salmonellosis and tuberculosis". It is U.S. law since 1944 that all raw-milk cheeses (including imports since 1951) must be aged at least 60 days. Australia has a wide ban on raw-milk cheeses as well, though in recent years exceptions have been made for Swiss Gruyère, Emmental and Sbrinz, and for French Roquefort. There is a trend for cheeses to be pasteurized even when not required by law.

Pregnant women may face an additional risk from cheese; the U.S. Centers for Disease Control has warned pregnant women against eating soft-ripened cheeses and blue-veined cheeses, due to the listeria risk, which can cause miscarriage or harm the fetus.

Cultural Attitudes

Although cheese is a vital source of nutrition in many regions of the world and is extensively consumed in others, its use is not universal.

A cheese merchant in a French market

A traditional Polish sheep's cheese market in Zakopane, Poland

Cheese is rarely found in East Asian cuisines, presumably for historical reasons. However, East Asian sentiment against cheese is not universal. In Nepal, the Dairy Development Corporation commercially manufactures cheese made from yak milk and a hard cheese made from either cow or yak milk knows as chhurpi. The national dish of Bhutan, ema datsi, is made from homemade yak or mare milk cheese and hot peppers. In Yunnan China, several ethnic minority groups produce Rushan and Rubing from cow's milk. Cheese consumption may be increasing in China, with annual sales dou-bling from 1996 to 2003 (to a still small 30 million U.S. dollars a year). Certain kinds of Chinese preserved bean curd are sometimes misleadingly referred to in English as "Chinese cheese" because of their texture and strong flavor.

Strict followers of the dietary laws of Islam and Judaism must avoid cheeses made with rennet from animals not slaughtered in a manner adhering to halal or kosher laws. Both faiths allow cheese made with vegetable-based rennet or with rennet made from animals that were processed in a halal or kosher manner. Many less orthodox Jews also believe that rennet undergoes enough processing to change its nature entirely and do not consider it to ever violate kosher law. As cheese is a dairy food, under kosher rules it cannot be eaten in the same meal with any meat.

Rennet derived from animal slaughter, and thus cheese made with animal-derived rennet, is not vegetarian. Most widely available vegetarian cheeses are made using rennet produced by fermentation of the fungus *Mucor miehei*. Vegans and other dairy-avoiding vegetarians do not eat conventional cheese, but some vegetable-based cheese substitutes (soy or almond) are used as substitutes.

Even in cultures with long cheese traditions, consumers may perceive some cheeses that are especially pungent-smelling or mold-bearing varieties such as Limburger or Roquefort, as unpalatable. Such cheeses are an acquired taste because they are processed using molds or microbiological cultures, allowing odor and flavor molecules to resemble those in rotten foods. One author stated: "An aversion to the odor of decay has the obvious biological value of steering us away from possible food poisoning, so it is no wonder that an animal food that gives off whiffs of shoes and soil and the stable takes some getting used to."

Collecting cheese labels is called "tyrosemiophilia".

Types of Cheese

Stilton, a blue cheese from England

There are several types of cheese, which are grouped or classified according to criteria such as length of ageing, texture, methods of making, fat content, animal milk, country or region of origin, etc. The method most commonly and traditionally used is based on moisture content, which is then further narrowed down by fat content and curing or ripening methods. The criteria may either be used singly or in combination, but with no single method being universally used. The combination of types produces around 500 different varieties recognised by the International Dairy Federation, over 400 identified by Walter and Hargrove, over 500 by Burkhalter, and over 1,000 by Sandine and Elliker. Some attempts have been made to rationalise the classification of cheese; a scheme was proposed by Pieter Walstra that uses the primary and secondary starter combined with moisture content, and Walter and Hargrove suggested classifying by production methods. This last scheme results in 18 types, which are then further grouped by moisture content.

Fresh, Whey, and Stretched Curd Cheeses

The main factor in categorizing these cheeses is age. Fresh cheeses without additional preservatives can spoil in a matter of days.

For these simplest cheeses, milk is curdled and drained, with little other processing. Examples include cottage cheese, cream cheese, curd cheese, farmer cheese, ca□, chhena, fromage blanc, queso fresco, paneer, and fresh goat's milk chèvre. Such cheeses are soft and spreadable, with a mild flavour.

Ricotta

Whey cheeses are fresh cheeses made from whey, a by-product from the process of producing other cheeses which would otherwise be discarded. Corsican brocciu, Italian ricotta, Romanian urda, Greek mizithra, Cypriot anari cheese and Norwegian Brunost are examples. Brocciu is mostly eaten fresh, and is as such a major ingredient in Corsican cuisine, but it can also be found in an aged form.

Traditional *pasta filata* cheeses such as Mozzarella also fall into the fresh cheese category. Fresh curds are stretched and kneaded in hot water to form a ball of Mozzarella, which in southern Italy is usually eaten within a few hours of being made. Stored in brine, it can easily be shipped, and it is known worldwide for its use on pizza.

Moisture: Soft to Hard

Categorizing cheeses by moisture content or firmness is a common but inexact practice. The lines between "soft", "semi-soft", "semi-hard", and "hard" are arbitrary, and many types of cheese are made in softer or firmer variants. The factor that controls cheese hardness is moisture content, which depends on the pressure with which it is packed into moulds, and upon aging time.

Emmentaler

Soft Cheese

Cream cheeses are not matured. Brie and Neufchâtel are soft-type cheeses that mature for more than a month.

Semi-soft Cheese

Semi-soft cheeses, and the sub-group *Monastery*, cheeses have a high moisture content and tend to be mild-tasting. Some well-known varieties include Havarti, Munster and Port Salut.

Medium-hard Cheese

Cheeses that range in texture from semi-soft to firm include Swiss-style cheeses such as Emmental and Gruyère. The same bacteria that give such cheeses their eyes also contribute to their aromatic and sharp flavours. Other semi-soft to firm cheeses include Gouda, Edam, Jarlsberg, Cantal, and Ca□caval. Cheeses of this type are ideal for melting and are often served on toast for quick snacks or simple meals.

Semi-hard or Hard Cheese

Harder cheeses have a lower moisture content than softer cheeses. They are generally packed into moulds under more pressure and aged for a longer time than the soft cheeses. Cheeses that are classified as semi-hard to hard include the familiar Cheddar, originating in the village of Cheddar in England but now used as a generic term for this style of cheese, of which varieties are imitated worldwide and are marketed by strength or the length of time they have been aged. Cheddar is one of a family of semi-hard or hard cheeses (including Cheshire and Gloucester), whose curd is cut, gently heated, piled, and stirred before being pressed into forms. Colby and Monterey Jack are similar but milder cheeses; their curd is rinsed before it is pressed, washing away some acidity and calcium. A similar curd-washing takes place when making the Dutch cheeses Edam and Gouda.

Hard cheeses — "grating cheeses" such as Parmesan and Pecorino Romano—are quite firmly packed into large forms and aged for months or years.

St. Pat Goat's Milk Cheese.

Source of Milk Used

Some cheeses are categorized by the source of the milk used to produce them or by the added fat content of the milk from which they are produced. While most of the world's commercially available cheese is made from cow's milk, many parts of the world also produce cheese from goats and sheep. Well-known examples include Roquefort (produced in France) and Pecorino (produced in Italy) from ewe's milk. One farm in Sweden also produces cheese from moose's milk. Sometimes cheeses marketed under the same name are made from milk of different animal—Feta style cheeses, for example, are made from sheep's milk in Greece and from cow's milk elsewhere.

Double cream cheeses are soft cheeses of cow's milk enriched with cream so that their FDM is 60–75% or, in the case of triple creams, at least 75%.

Mold

Vacherin du Haut-Doubs cheese, a French cheese with a white *Penicillium* mold rind.

There are three main categories of cheese in which the presence of mold is an important feature: soft ripened cheeses, washed rind cheeses and blue cheeses.

Soft-ripened

Soft-ripened cheeses begin firm and rather chalky in texture, but are aged from the exterior inwards by exposing them to mold. The mold may be a velvety bloom of *P. camemberti* that forms a flexible white crust and contributes to the smooth, runny, or gooey textures and more intense flavors of these aged cheeses. Brie and Camembert, the most famous of these cheeses, are made by allowing white mold to grow on the outside of a soft cheese for a few days or weeks. Goat's milk cheeses are often treated in a similar manner, sometimes with white molds (Chèvre-Boîte) and sometimes with blue.

Washed-rind

Washed-rind cheeses are soft in character and ripen inwards like those with white molds; however, they are treated differently. Washed-rind cheeses are periodically cured in a solution of saltwater brine and/or mold-bearing agents that may include beer, wine, brandy and spices, making their surfaces amenable to a class of bacteria *Brevibacterium linens* (the reddish-orange "smear bacteria") that impart pungent odors and distinctive flavors, and produce a firm, flavorful rind around the cheese. Washed-rind cheeses can be soft (Limburger), semi-hard, or hard (Appenzeller). The same bacteria can also have some impact on cheeses that are simply ripened in humid conditions, like Camembert. The process requires regular washings, particularly in the early stages of production, making it quite labor-intensive compared to other methods of cheese production.

Smear-ripened

Some washed-rind cheeses are also smear-ripened with solutions of bacteria or fungi, most commonly *Brevibacterium linens*, *Debaryomyces hansenii*, and/or *Geotrichum candidum*) which usually gives them a stronger flavor as the cheese matures. In some cases, older cheeses are smeared on young cheeses to transfer the microorganisms. Many, but not all, of these cheeses have a distinctive pinkish or orange coloring of the exterior. Unlike with other washed-rind cheeses, the washing is done to ensure uniform growth of desired bacteria or fungi and to prevent the growth of undesired molds. Notable examples of smear-ripened cheeses include Munster and Port Salut.

Blue

So-called blue cheese is created by inoculating a cheese with *Penicillium roqueforti* or *Penicillium glaucum*. This is done while the cheese is still in the form of loosely pressed curds, and may be further enhanced by piercing a ripening block of cheese with skewers in an atmosphere in which the mold is prevalent. The mold grows within the cheese as it ages. These cheeses have distinct blue veins, which gives them their name and, often, assertive flavors. The molds range from pale green to dark blue, and may be accompanied by white and crusty brown molds. Their texture can be soft or firm. Some of the most renowned cheeses are of this type, each with its own distinctive color, flavor, texture and aroma. They include Roquefort, Gorgonzola and Stilton.

Brined

Feta, a brined curd cheese

Brined or pickled cheese is matured in a solution of brine in an airtight or semi-permeable container. This process gives the cheese good stability, inhibiting bacterial growth even in hot countries. Brined cheeses may be soft or hard, varying in moisture content, and in colour and flavour, according to the type of milk used; though all will be rindless, and generally taste clean, salty and acidic when fresh, developing some piquancy when aged, and most will be white. Varieties of brined cheese include bryndza, feta, halloumi, and sirene. Brined cheese is the main type of cheese produced and eaten in the Middle East and Mediterranean areas.

Processed

Processed cheese is made from traditional cheese and emulsifying salts, often with the addition of milk, more salt, preservatives, and food colouring. Its texture is consistent, and melts smoothly. It is sold packaged and either pre-sliced or unsliced, in several varieties. Some are sold as sausage-like logs and chipolatas (mostly in Germany and USA), and some are moulded into the shape of animals and objects. It is also available in aerosol cans in some countries.

Some, if not most, varieties of processed cheese are made using a combination of real cheese waste (which is steam-cleaned, boiled and further processed) whey powders, and various mixtures of vegetable and/or palm oils and fats. Some processed "cheese" slices contain as little as 2-6% cheese; some have smoke flavours added.

Cottage Cheese

Cottage cheese is a fresh cheese curd product with a mild flavor.

A tub of cottage cheese

Homemade cottage cheese

Origin

The first known use of the term "cottage cheese" dates back to 1831 and is believed to have originated because the simple cheese was usually made in cottages from any milk left over after making butter. The curds and whey from the Little Miss Muffet nursery rhyme is another dish made from curds with whey, but it is uncertain what their consistency was, if they were drained at all or how they were curdled (which affects the flavor). Some writers claim they are equivalent or similar.

Manufacture

It is drained, but not pressed, so some whey remains and the individual curds remain loose. The curd is usually washed to remove acidity, giving sweet curd cheese. It is not aged or colored. Different styles of cottage cheese are made from milks with different fat levels and in small-curd or large-curd preparations. Cottage cheese which is pressed becomes hoop cheese, farmer cheese, pot cheese, or queso blanco.

Curd Size

The curd size is the size of the chunks in the cottage cheese. The two major types of cottage cheese are small-curd, high-acid cheese made without rennet, and large-curd, low-acid cheese made with rennet. Rennet is a natural complex of enzymes that speeds curdling and keeps the curd that forms from breaking up; adding it shortens the cheesemaking process, resulting in a lower acid and larger curd cheese, and reduces the amount of curd poured off with leftover liquid (the whey). Sometimes large-curd cottage cheese is called "chunk style."

Consumption

Cottage cheese can be eaten in a variety of different ways: by itself, with fruit and sugar, with salt and pepper, with fruit puree, on toast, with tomatoes, with granola and cinnamon, in salads, as a chip dip, as a replacement for mayonnaise in tuna salad or used as an ingredient in recipes such as jello salad and various desserts. Cottage cheese with fruit such as pears, peaches, or mandarin oranges is a standard side dish in many "home cooking" or meat-and-three restaurants' menus in the United States.

Nutrition

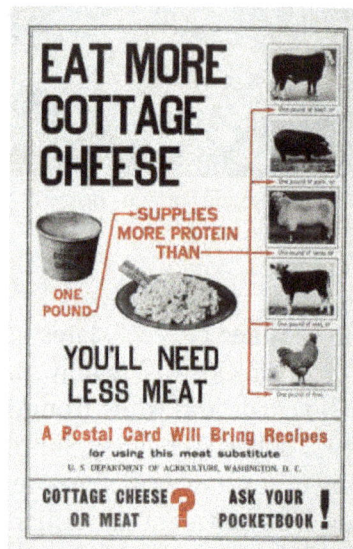

World War I poster encouraging U.S. citizens to consume cottage cheese as an alternative to meat products.

A 113-g (4-oz) serving of 4% fat product has about 120 calories, 5 g fat (3 g saturated), 3 g carbohydrates, and 12 g protein. It also contains about 500 mg sodium, 70 mg calcium, and 20 mg cholesterol.

Some manufacturers also produce low-fat and nonfat varieties. A fat-free kind of a similar serving size has 80 calories, 0 g fat (0 g saturated), 6 g carbohydrates, and 14 g protein.

Cottage cheese is popular among dieters and some health food devotees. It is a favorite food among bodybuilders, runners, swimmers and weightlifters for its high content

Cheesemaking

of casein protein (a longer-lasting protein) while being relatively low in fat. Pregnant women are advised that cottage cheese is safe to eat, whereas some cheese products are not recommended during pregnancy.

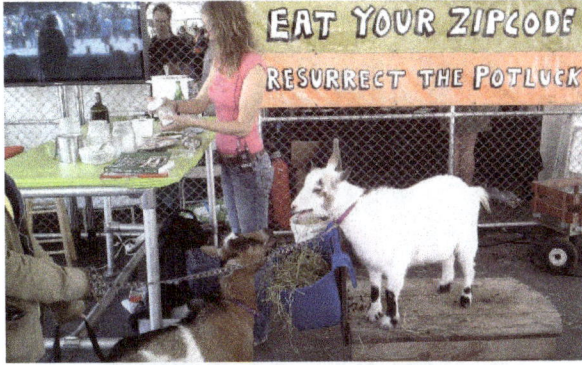

A cheesemaking workshop with goats at Maker Faire 2011. The sign declares, "Eat your Zipcode", in reference to the locavore movement

During industrial production of Emmental cheese, the as-yet-undrained curd is broken by rotating mixers.

The production of cheese, like many other food preservation processes allows the nutritional and economic value of a food material, in this case milk, to be preserved. It allows the consumer to choose (within limits) when to consume the food rather than have to consume it straight away, and it allows the product to be altered which gives it higher value.

Cheesemaking may have originated from nomadic herdsmen who stored milk in vessels made from the sheep's and goats' stomachs. Because their stomach linings contains a mix of lactic acid, wild bacteria as milk contaminants and rennet, the milk would ferment and coagulate. A product reminiscent of yogurt would have been produced, which, through gentle agitation and the separation of curds from whey would have resulted in the production of cheese; the cheese being essentially a concentration of the major milk protein, casein, and milk fat. The whey proteins, other minor milk proteins, and the lactose are all removed in the cheese whey.

Process

The production of Gruyère cheese at the cheesemaking factory of Gruyères, Canton of Fribourg, Switzerland

The job of the cheesemaker is to control the spoiling of milk into cheese. The milk may be from a cow, goat, sheep or buffalo, although worldwide cow's milk is most commonly used. The cheesemaker applies craft and skill to the practise of cheesemaking, intending to produce a product with specific characteristics and organoleptic requirements (appearance, aroma, taste, texture) that are consistent every time it is made. This is not to say, of course, there is no room for variety or innovation, but a particular cheese needs to be made a particular way. Thus, the crafts and skills employed by the cheesemaker to make a Camembert will be similar to, but not quite the same as, those used to make Cheddar.

In modern industrial cheesemaking factories (sometimes called creameries) the craft elements of cheesemaking are retained to some extent, but there is more science than craft. This is seen particularly in factories that use computer-aided manufacturing. The end product is very predictable. So in contrast, individual cheesemakers tend to operate on a much smaller scale and sell "handmade" products; each batch may differ, but their customers expect and anticipate this, much like with wines, teas and many other natural products.

Some cheeses may be deliberately left to ferment from naturally airborne spores and bacteria; this generally leads to a less consistent product but one that is highly valuable in a niche market for exactly that reason, no two are ever quite the same.

Culturing

To make cheese the cheesemaker brings milk (possibly pasteurised) in the cheese vat to a temperature required to promote the growth of the bacteria that feed on lactose and thus ferment the lactose into lactic acid. These bacteria in the milk may be wild, as is the case with unpasteurised milk, added from a culture, frozen or freeze dried concentrate of starter bacteria. Bacteria which produce only lactic acid during fermentation are homofermentative; those that also produce lactic acid and other compounds such as carbon dioxide, alcohol, aldehydes and ketones are heterofermentative. Fermentation using homofermentative bacteria is important in the production of cheeses such as Cheddar, where a clean, acid flavour is required. For cheeses such as Emmental the use of heterofermentative bacteria is necessary to produce the compounds that give characteristic fruity flavours and, importantly, the gas that results in the formation of bubbles in the cheese ('eye holes').

Cheesemakers choose starter cultures to give a cheese its specific characteristics. Also, if the cheesemaker intends to make a mould-ripened cheese such as Stilton, Roquefort or Camembert, mould spores (fungal spores) may be added to the milk in the cheese vat or can be added later to the cheese curd.

Coagulation

When during the fermentation the cheesemaker has gauged that sufficient lactic acid has been developed, rennet is added to cause the casein to precipitate. Rennet contains the enzyme chymosin which converts κ-casein to para-κ-caseinate (the main component of cheese curd, which is a salt of one fragment of the casein) and glycomacropeptide, which is lost in the cheese whey. As the curd is formed, milk fat is trapped in a casein matrix. After adding the rennet, the cheese milk is left to form curds over a period of time.

Draining

Fresh chevre hanging in cheesecloth to drain.

Once the cheese curd is judged to be ready, the cheese whey must be released. As with many foods the presence of water and the bacteria in it encourages decomposition. The cheesemaker must, therefore, remove most of the water (whey) from the cheese milk, and hence cheese curd, to make a partial dehydration of the curd. This ensures a product of good quality and that will keep. There are several ways to separate the curd from the whey, and it is again controlled by the cheesemaker.

Scalding

If making Cheddar (or many other hard cheeses) the curd is cut into small cubes and the temperature is raised to around 39 °C (102 °F) to 'scald' the curd particles. Syneresis occurs and cheese whey is expressed from the particles. The Cheddar curds and whey are often transferred from the cheese vat to a cooling table which contains screens that allow the whey to drain, but which trap the curd. The curd is cut using long, blunt knives and 'blocked' (stacked, cut and turned) by the cheesemaker to promote the release of cheese whey in a process known as 'cheddaring'. During this process the acidity of the curd increases and when the cheesemaker is satisfied it has reached the required level, around 0.65%, the curd is milled into ribbon shaped pieces and salt is mixed into it to arrest acid development. The salted green cheese curd is put into cheese moulds lined with cheesecloths and pressed overnight to allow the curd particles to bind together. The pressed blocks of cheese are then removed from the cheese moulds and are either bound with muslin-like cloth, or waxed or vacuum packed in plastic bags to be stored for maturation. Vacuum packing removes oxygen and prevents mould (fungal) growth during maturation, which depending on the wanted final product may be a desirable characteristic or not.

Mould-ripening

In contrast to cheddaring, making cheeses like Camembert requires a more gentle treatment of the curd. It is carefully transferred to cheese hoops and the whey is allowed to drain from the curd by gravity, generally overnight. The cheese curds are then removed from the hoops to be brined by immersion in a saturated salt solution. The salt absorption stops bacteria growing, as with Cheddar. If white mould spores have not been added to the cheese milk the cheesemaker applies them to the cheese either by spraying the cheese with a suspension of mould spores in water or by immersing the cheese in a bath containing spores of, e.g., *Penicillium candida*.

By taking the cheese through a series of maturation stages where temperature and relative humidity are carefully controlled, the cheesemaker allows the surface mould to grow and the mould-ripening of the cheese by fungi to occur. Mould-ripened cheeses ripen very quickly compared to hard cheeses (weeks against months or years). This is because the fungi used are biochemically very active when compared with starter bacteria. Some cheeses are surface-ripened by moulds, such as Camembert and Brie, some are ripened internally, such as Stilton, which is pierced by the cheesemaker with stainless steel wires, to admit air to promote mould spore germination and growth, as with *Penicillium roqueforti*. Surface ripening of some cheeses, such as Saint-Nectaire, may also be influenced by yeasts which contribute flavour and coat texture. Others are allowed by the cheesemaker to develop bacterial surface growths which give characteristic colours and appearances, e.g. by the growth of *Brevibacterium linens* which gives an orange coat to cheeses.

Quality Control

Cheesemakers must be skilled in the grading of cheese to assess quality, defects and suitability for release from the maturing store for sale. The grading process is one of sampling by sight, smell, taste and texture. Part of the cheesemaker's skill lies in the ability to predict when a cheese will be ready for sale or consumption, as the characteristics of cheese change constantly during maturation.

A cheesemaker is thus a person who has developed the knowledge and skills required to convert milk into cheese, by controlling precisely the types and amounts of ingredients used, and the parameters of the cheesemaking process, to make specific types and qualities of cheese. Most cheesemakers by virtue of their knowledge and experience are adept at making particular types of cheese. Few, if any, could quickly turn their hand to making other kinds. Such is the specialisation of cheesemaking.

Making artisan cheese or farmstead cheese in the United States has become more popular in recent times, as an extension of the craft of cheesemaking.

Cheesecloth

Cheesecloth is a loose-woven gauze-like cotton cloth used primarily in cheese making and cooking.

Cheese hung in cheesecloth to strain whey as part of cheesemaking

Grades

Cheesecloth is available in at least seven different grades, from open to extra-fine weave. Grades are distinguished by the number of threads per inch in each direction.

Grade	Vertical × horizontal threads per inch	Vertical x horizontal threads/cm
#10	20 × 12	8 x 5
#40	24 × 20	9.5 x 8
#50	28 × 24	11 x 9.5
#60	32 × 28	12.5 x 11
#90	44 × 36	17.5 x 14

Uses

Food Preparation

The primary use of cheesecloth is in some styles of cheesemaking, where it is used to remove whey from cheese curds, and to help hold the curds together as the cheese is formed. Cheesecloth is also used in straining stocks and custards, bundling herbs, making tofu and ghee, and thickening yogurt. Queso blanco and queso fresco are Spanish and Mexican cheeses that are made from whole milk using cheesecloth. Quark is a type of German unsalted cheese that is sometimes formed with cheesecloth. Paneer is a kind of Indian fresh cheese that is commonly made with cheesecloth. Fruitcake is wrapped in rum-infused cheesecloth during the process of "feeding" the fruitcake as it ripens.

Other Uses

Cheesecloth can also be used for several printmaking processes including lithography for wiping up gum arabic. In intaglio a heavily starched cheesecloth called tarlatan is used for wiping away excess ink from the printing surface.

Cheesecloth #60 is used in product safety and regulatory testing for potential fire hazards. Cheesecloth is wrapped tightly over the device under test, which is then subjected to simulated conditions such as lightning surges conducted through power or telecom cables, power faults, etc. The device may be destroyed but must not ignite the cheesecloth. This is to ensure that the device can fail safely, and not start electrical fires in the vicinity.

Cheesecloth made to United States Federal Standard CCC-C-440 is used to test the durability of optical coatings per United States Military Standard MIL-C-48497. The optics are exposed to a 95%-100% humidity environment at 120 °F (49 °C) for 24 hours, and then a $\frac{1}{4}$ inch (6.4 mm) thick by $\frac{3}{8}$ in (9.5 mm) wide pad of cheese cloth is rubbed over the optical surface for at least 50 strokes under at least 1 pound-force (4.4 N). The optical surface is examined for streaks or scratches, and then its optical performance is measured to ensure that no deterioration occurred.

Cheesecloth is used in India and Pakistan for making summer shirts. Cheesecloth material shirts were popular for beachwear during the 1960s and 1970s in the United States. Cheesecloth has been used to create the illusion of "ectoplasm" during spirit channelings or other ghost related phenomena.

Cheese Ripening

The effect of dairy salt in Cheddar cheese making: increased use of salt reduces moisture and slows the ripening process.

Cheese ripening, alternatively cheese maturation of affinage, is a process in cheesemaking. It is responsible for the distinct flavour of cheese, and through the modification of *"ripening agents"*, determines the features that define many different varieties of cheeses, such as taste, texture, and body. The process is "characterized by a series of complex physical, chemical and microbiological changes" that incorporates the agents of: "bacteria and enzymes of the milk, lactic culture, rennet, lipases, added moulds or yeasts, and environmental contaminants." The majority of cheese is ripened, save for fresh cheese.

History

Cheese ripening was not always the highly industrialised process it is today; in the past, cellars and caves were used to ripen cheeses instead of the current highly regulated process involving machinery and biochemistry. Some cheeses still are made using more historical methods, such as the blue cheese Roquefort, which is required to be ripened in designated caves in south-eastern France. However, with the invention of refrigeration in the 20th century, the process evolved considerably, and is much more efficient at producing a consistent quality of cheese, at a faster pace, and a lower cost (depending on the type of cheese).

Process

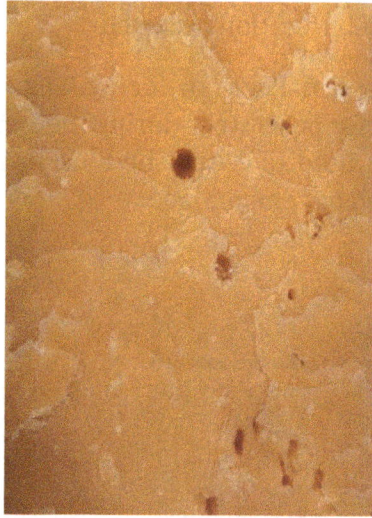

Extra aged Beemster cheese, 26+ months old. Note the salt crystals.

After the initial manufacturing process of the cheese is done, the cheese ripening process occurs. This process is especially important, since it defines the flavour and texture of the cheese, which differentiates the many varieties. Duration is dependent on the type of cheese and the desired quality, but "three weeks to two or more years" is the general requirement for most cheeses.

Ripening is influenced by a variety of factors, ranging from the microflora to the curd, and others. The enzymatic process is the most crucial process for all cheeses, although bacteria plays a role in many varieties. The most important agents in this process include the four following elements: "Rennet, or a substitute for rennet, starter bacteria and associated enzymes, milk enzymes, second starter bacteria and associated enzymes, and non-starter bacteria". Each of these factors affects the cheese-ripening process differently, and has been the subject of much research. It is important for manufacturers to understand how each of these elements work, so that they are able to maintain the quality of the cheese while producing the cheese at an acceptable investment of time and cost. These agents contribute to the three primary reactions that define cheese ripening: glycolysis, proteolysis, and lipolysis.

By taking the cheese through a series of maturation stages where temperature and relative humidity are carefully controlled, the cheese maker allows the surface mould to grow and the mould ripening of the cheese by fungi to occur. Mould-ripened cheeses ripen faster than hard cheeses, in weeks as opposed to the typical months or even years. This is because the fungi used are more biochemically active than the starter bacteria. Where the ripening occurs is largely dependent on the type of cheese: some cheeses are surface-ripened by moulds, such as Camembert and Brie; and some are ripened internally, such as Stilton. Surface ripening of some cheeses, such as Saint-Nectaire cheese, may also be influenced by yeasts which contribute flavour and coat texture. Others are allowed by the cheesemaker to develop bacterial surface growths which give characteristic colours and appearances. The growth of *Brevibacterium linens*, for example, creates an orange coat to cheeses.

In contrast to cheddaring, making cheeses like Camembert requires a more gentle treatment of the curd. It is carefully transferred to cheese hoops and the whey is allowed to drain from the curd by gravity, generally overnight. The cheese curds are then removed from the hoops to be brined by immersion in a saturated salt solution. This is because the amount of salt has a large effect on the rate of proteolysis in the cheese, stopping the bacteria from growing. If white-mould spores have not been added to the cheese milk, the cheese maker applies them to the cheese either by spraying the cheese with a suspension of mould spores in water, or by immersing the cheese in a bath containing spores of, e.g., *Penicillium candida*.

Effect on Features

Eyes

Emmental with eyes. Emmental tastes sweeter due to proline.

The round holes that are a characteristic feature of Swiss-type cheese (e.g. Emmentaler cheese) and some Dutch-type cheeses are called "eyes". They are bubbles of carbon dioxide that is produced by bacteria in the cheese.

In Swiss-type cheeses, the eyes form as a result of the activity of propionic acid bacteria (*propionibacteria*), notably *Propionibacterium freudenreichii* subsp. *shermanii*. In Dutch-type cheeses, the CO_2 that forms the eyes results from the metabolisation of citrate by citrate-positive ("Cit+") strains of lactococci.

Taste

The process of cheese ripening affects the taste of the final product. If the product is not ripened, the resulting cheese is tasteless, and so, all cheese is ripened except for fresh cheeses. Different factors define taste in cheese, including casein, fat, brine, and many other elements. Brine, as an example, mixes with saliva, delivering the flavour of the cheese to the taste buds and determining the cheese's moistness. Many of these elements are specific to the type of cheese. For instance, proline is more abundant in Emmental than in any other type of cheese, and gives the cheese its much sweeter taste.

References

- Patrick F. Fox; P. F. Fox (2000). Fundamentals of cheese science. Springer, 2000. p. 388. ISBN 9780834212602. Retrieved March 21, 2011.

- Patrick F. Fox; P. F. Fox (1999-02-28). Cheese: chemistry, physics and microbiology, Volume 1. Springer, 1999. p. 1. ISBN 9780834213388. Retrieved March 23, 2011.

- Barbara Ensrud, (1981) The Pocket Guide to Cheese, Lansdowne Press/Quarto Marketing Ltd., ISBN 0-7018-1483-7

- McGee, Harold (2004). On Food and Cooking: The Science and Lore of the Kitchen. Scribner. ISBN 9780684800011.

- Barbara Ensrud, (1981) The Pocket Guide to Cheese, Lansdowne Press/Quarto Marketing Ltd., ISBN 0-7018-1483-7

- Driscoll, Michael; Meredith Hamiltion; Marie Coons (May 2003). A Child's Introduction Poetry. 151 West 19th Street New York, NY 10011: Black Dog & Leventhal Publishers. p. 10. ISBN 1-57912-282-5.

- Fox, P.F. (ed.). Cheese: Chemistry, Physics, and Microbiology, Volume 1: General Aspects. Academic Press. ISBN 978-0-12-263652-3.

- Fox, Patrick. Cogan, Timothy. Guinee, Timothy. Fundamentals of Cheese Science. Springer Publishing. ISBN 978-0-8342-1260-2.

- Workman D (12 April 2016). "Cheese Exports by Country in 2015". World's Top Exports. Retrieved 2 June 2016.

- "Nutrition facts for various cheeses per 100 g". Nutritiondata.com. Conde Nast; republished from the USDA National Nutrient Database, version SR-21. 2014. Retrieved 1 June 2016.

- "World production of cheese (all kinds) in 2013; Browse Data/Livestock Processed/World". United Nations Food and Agriculture Organization, Statistics Division (FAOSTAT). 2015. Retrieved 2 June 2016.

Other Fermented Dairy Products

Fermented milk products are the products that are fermented by using lactic acid bacteria such as lactobacillus and leuconostoc. The production of milk products can be traced back to around 10,000 BC. The products discussed in this section are filmjölk, kumis, kefir, chal, lassi and yakult. The main products of dairy products are discussed in the following chapter.

Fermented Milk Products

Fermented milk products, also known as cultured dairy foods, cultured dairy products, or cultured milk products, are dairy foods that have been fermented with lactic acid bacteria such as Lactobacillus, Lactococcus, and Leuconostoc. The fermentation process increases the shelf-life of the product, while enhancing the taste and improving the digestibility of milk. There is evidence that fermented milk products have been produced since around 10,000 BC. A range of different Lactobacilli strains has been grown in laboratories allowing for a wide range of cultured milk products with different tastes.

Dadiah is a traditional fermented milk of West Sumatra, Indonesia prepared with fresh, raw and unheated buffalo milk

Products

Many different types of cultured milk products can be found around the world.

Soured Milk

Country/region of origin	Product(s)
	acidophilus milk
	buttermilk
	cheese
	curd
Armenia	matzoon
Arab World	leben, kishk
Central Asia	chal/shubat, chalap, kumis, qatyq, qurt, suzma
Brittany	laezh-ribod
Bulgaria	kiselo mlyako
Czech Republic	kefir or Acidofilni mleko
Denmark	kærnemælk, tykmælk, and ymer
Dominican Republic	Boruga
Estonia	soured milk and kefir
Finland	piimä and viili
Germany	Sauermilch or Dickmilch (soured milk or thickened milk), Quark
Georgia	matsoni
Greece	Xynogalo or Xynogala
Hungary	aludttej or yogurt or kefir
Iceland	skyr and súrmjólk
India	paneer, dahi, lassi, chaas, mattha, mishti doi and shrikhand
Indonesia	dadiah
Iran	doogh, kashk, ghara
Middle East	leben
Japan	calpis
Latvia	rūgušpiens, kefīrs, skābputra
Lithuania	rūgpienis, kefyras
Macedonia	kiselo mleko
Mexico	jocoque
Mongolia	airag, byaslag, tarag, khuruud
Netherlands	karnemelk (buttermilk)

Nicaragua	leche agria (soured milk)
Norway	surmelk or kulturmelk, kefir, and tjukkmjølk
Pakistan	dahi and lassi
Poland	soured milk (including "acidofilne" milk), kefir, buttermilk, twaróg
Romania	lapte bătut, lapte acru, kefir and sana
Russia, Ukraine, Belarus	kefir, prostokvasha, ryazhenka, varenets, tvorog
Rwanda	kivuguto
Scotland	blaand
Serbia	kiselo mleko and yogurt
Slovakia	kefir or acidofilne mlieko
Slovenia	kislo mleko
South Africa	amasi ("maas" in Afrikaans)
Sweden	filmjölk, långfil and A-fil (fil is the short form of filmjölk)
Turkic countries	ayran, qatiq
United States	clabber
Bosnia and Herzegovina	kiselo mlijeko and kefir
Zambia	Mabisi
Zimbabwe	lacto
Burundi	urubu
Kenya	Kule Naoto, Maziwa Lala, Mursik
Ethiopia	ergo
Sudan	rob

Soured Cream

Country/region of origin	Product(s)
	cheese
	sour cream
Central Asia	kaymak
Central & Eastern Europe	smetana
Croatia	mileram/kiselo vrhnje
Finland	kermaviili
France	crème fraîche

Iceland	sýrður rjómi
Hungary	tejföl
Latvia	skābais krējums
Lithuania	grietinė
Mexico	crema/cream espesa
Norway	rømme
Romania	smântână
Serbia	kisela pavlaka
Sweden	gräddfil

Comparison chart

Product	Alternative names	Typical milkfat content	Typical shelf life at 4 °C	Fermentation agent	Description
Cheese		1-75%	varies	a variety of bacteria and/or mold	Any number of solid fermented milk products.
Crème fraîche	creme fraiche	30-40%	10 days	naturally occurring lactic acid bacteria in cream	Mesophilic fermented cream, originally from France; higher-fat variant of sour cream
Cultured sour cream	sour cream	14–40%	4 weeks	Lactococcus lactis subsp. lactis*	Mesophilic fermented pasteurized cream with an acidity of at least 0.5%. Rennet extract may be added to make a thicker product. Lower fat variant of crème fraîche
Filmjölk	fil	0.1-4.5%	10–14 days	Lactococcus lactis* and Leuconostoc	Mesophilic fermented milk, originally from Scandinavia
Yogurt	yoghurt, yogourt, yoghourt	0.5–4%	35–40 days	Lactobacillus bulgaricus and Streptococcus thermophilus	Thermophilic fermented milk, cultured with Lactobacillus bulgaricus and Streptococcus thermophilus

Kefir	kephir, kewra, talai, mudu kekiya, milkkefir, búlgaros	0-4%	10–14 days	Kefir grains, a mixture of bacteria and yeasts	A fermented beverage, originally from the Caucasus region, made with kefir grains; can be made with any sugary liquid, such as milk from mammals, soy milk, or fruit juices
Kumis	koumiss, kumiss, kymys, kymyz, airag, chigee	4%?	10–14 days	Lactobacilli and yeasts	A carbonated fermented milk beverage traditionally made from horse milk
Viili	filbunke	0.1-3.5%	14 days	Lactococcus lactis subsp. cremoris, Lactococcus lactis* biovar. diacetylactis, Leuconostoc mesenteroides subsp. cremoris and Geotrichum candidum	Mesophilic fermented milk that may or may not contain fungus on the surface; originally from Sweden; a Finnish specialty
Cultured buttermilk		1–2%	10 days	Lactococcus lactis* (Lactococcus lactis subsp. lactis*, Lactococcus lactis subsp. cremoris, Lactococcus lactis biovar. diacetylactis and Leuconostoc mesenteroides subsp. cremoris)	Mesophilic fermented pasteurized milk
Acidophilus milk	acidophilus cultured milk	0.5-2%	2 weeks	Lactobacillus acidophilus	Thermophilic fermented milk, often lowfat (2%, 1.5%) or nonfat (0.5%), cultured with Lactobacillus acidophilus

* Streptococcus lactis has been renamed to Lactococcus lactis subsp. lactis

Filmjölk

Filmjölk in a glass.

Filmjölk, also known as fil, is a traditional fermented milk product from Sweden, and a common dairy product within the Nordic countries. It is made by fermenting cow's milk with a variety of bacteria from the species *Lactococcus lactis* and *Leuconostoc mesenteroides*. The bacteria metabolize lactose, the sugar naturally found in milk, into lactic acid which means people who are lactose intolerant can tolerate it better than other dairy products. The acid gives filmjölk a sour taste and causes proteins in the milk, mainly casein, to coagulate, thus thickening the final product. The bacteria also produce a limited amount of diacetyl, a compound with a buttery flavor, which gives filmjölk its characteristic taste.

Filmjölk is similar to cultured buttermilk or kefir in consistency and has a mild and slightly acidic taste. It has a shelf-life of around 10–14 days at refrigeration temperature.

Overview

In the Nordic countries, filmjölk is often eaten with breakfast cereal, muesli or crushed crisp bread on top. Some people add sugar, jam, apple sauce, cinnamon, ginger, fruits, and/or berries for extra taste. In Norwegian it is called surmelk (Nynorsk: surmjølk) (sourmilk) but the official name is kulturmelk (Nynorsk: kulturmjølk). The drink is also popular in Latvian kitchens, where it is called rūgušpiens, rūgtpiens (fermented milk or sourmilk) and can be bought ready from stores but is more commonly made at home, additionally it can also be purchased and is popular in the neighboring country, Lithuania, where it is called rūgpienis or raugintas pienas (sour/fermented milk), due to its popularity it can be bought in majority of the stores alongside kefir.

Manufactured filmjölk is made from pasteurised, homogenised, and standardised cow's milk. Although homemade filmjölk has been around for a long time (written records from the 18th century speak of filmjölk-like products, but it has probably been around since the Viking Age or longer), it was first introduced to the Swedish market as a consumer product in 1931 by the Swedish dairy cooperative Arla. The first filmjölk was unflavoured and contained 3% milkfat. Since the 1960s, different varieties of unflavoured filmjölk have been marketed in Swedish grocery stores. Långfil, a more elastic variant of filmjölk was introduced in 1965; lättfil, filmjölk with 0.5% milkfat was introduced in 1967; and mellanfil, filmjölk with 1.5% milkfat was introduced in 1990. In 1997, Arla introduced its first flavoured filmjölk: strawberry flavoured filmjölk. The flavoured filmjölk was so popular that different flavours soon followed. By 2001, almost one third of the filmjölk sold in Sweden was flavoured filmjölk. Since 2007, variations of filmjölk include filmjölk with various fat content, filmjölk flavoured with fruit, vanilla, or honey, as well as filmjölk with probiotic bacteria that is claimed to be extra healthful, such as Onaka fil which contains Bifidobacterium lactis (a strain of bacteria popular in Japan) and Verum Hälsofil which contains *Lactococcus lactis* L1A in quantities of at least 10 billion live bacteria per deciliter.

In English

There is no single accepted English term for fil or filmjölk. In the United States it is referred to as 'long milk'. Fil and/or filmjölk has been translated to English as *sour milk*, *soured milk*, *acidulated milk*, *fermented milk*, and *curdled milk*, all of which are nearly synonymous and describe filmjölk but do not differentiate filmjölk from other types of soured/fermented milk. Filmjölk has also been described as *viscous fermented milk* and *viscous mesophilic fermented milk*,. Furthermore, articles written in English can be found that use the Swedish term *filmjölk*, as well as the Anglicised spellings *filmjolk*, *fil mjölk*, and *fil mjolk*.

In baking, when filmjölk is called for, cultured buttermilk can be substituted.

In Finland Swedish

In Finland Swedish, the dialects of Swedish spoken by Swedish-speaking Finns, *fil* in Finland is the equivalent of *filbunke* in Sweden. Not all variants of filmjölk are found in Finland, normally only *filbunke* and *långfil*. Swedish-speaking Finns usually use the word *surmjölk*, which is the older name for *filmjölk* (also in Sweden) or *piimä* (in Finnish), which is a fermented milk product that is thinner than filmjölk and resembles cultured buttermilk.

Types of filmjölk in Sweden

In Sweden, there are five Swedish dairy cooperatives that produce filmjölk: Arla Foods, Falköpings Mejeri, Gefleortens Mejeri, Norrmejerier, and Skånemejerier. In addition, Wapnö AB, a Swedish dairy company, and Valio, a Finnish dairy company, also sell a limited variety of filmjölk in Sweden. Prior to the manufacture of filmjölk, many families made filmjölk at home.

Fil culture is a variety of bacterium from the species *Lactococcus lactis* and *Leuconostoc mesenteroides*, e.g., Arla's fil culture contains *Lactococcus lactis* subsp. *lactis*, *Lactococcus lactis* subsp. *cremoris*, *Lactococcus lactis* biovar. *diacetylactis*, and *Leuconostoc mesenteroides* subsp. *cremoris*.

Classic Filmjölk Variants

Name	Literal translation	Milkfat content	Fermentation culture	Produced by	Year introduced	Description
Filmjölk		2.5%-3%	fil culture	Arla Foods, Falköpings Mejeri, Gefleortens Mejeri, Milko, Norrmejerier, Skånemejerier, Wapnö AB	1931 (Arla)	"Regular" filmjölk. Filmjölk made from 3% milkfat. Comes unflavoured and flavoured. Also comes in a variant made from organic milk, a low-lactose variant that has been treated with lactase enzyme, a variant with added fiber (f-fil, fil med fiber), and a variant with higher milkfat content (Arla Vår finaste filmjölk, 3.8–4.5% milkfat). Has been in the Swedish language since 1741.
Mellanfil	middle (lowfat) filmjölk	1.3%, 1.5%	fil culture	Arla Foods, Falköpings Mejeri, Gefleortens Mejeri, Milko, Norrmejerier, Skånemejerier	1990 (Arla)	Filmjölk made from 1.5% milkfat. Comes unflavoured only.
Lättfil	light (nonfat) filmjölk	0.4%, 0.5%	fil culture	Arla Foods, Falköpings Mejeri, Gefleortens Mejeri, Milko, Norrmejerier, Skånemejerier, Wapnö AB	1967 (Arla), 1968	Filmjölk made from 0.5% milkfat. Comes unflavoured and flavoured. Also comes in a low-lactose variant that has been treated with lactase enzyme.

Långfil *fi:* pit-käviili	long fil	3%	fil culture + *Lactococcus lactis* subsp. *lactis* var. *longi*	Arla Foods, Gefleortens Mejeri, Norrmejerier	1965 (Arla)	Filmjölk with a characteristic long and almost elastic texture due to *Lactococcus lactis* var. *longi*, a strain of bacteria that converts the carbohydrates in milk into long chains of polysaccharides. Comes unflavoured only. More common in northern Sweden. Sometimes eaten with ground ginger. Has been in the Swedish language since 1896.
Bollnäsfil	Bollnäs fil	3%	fil culture from Bollnäs	Milko		Filmjölk that originated in Bollnäs. Comes unflavoured or vanilla flavoured.
Fjällfil	fell fil	0.8%, 3.8–4.5%	special fil culture	Milko		Available as unflavoured, with birch sap, or raspberry.
Filbunke *fi-se:* Fil *fi:* Viili	bowl of fil	1%, 1.9%, 2.2%, 2.5%, 3%, 3.5%, 4%	special fil culture	Milko, Valio		Milk that has fermented, unstirred, in small bowls. Has a pudding-like consistency. Similar to unstirred långfil. Traditionally made in small bowls from (unpasteurized and unhomogenized) raw milk, which normally contains some cream. The cream forms a yellowish layer of sour cream on top. Comes unflavoured and flavoured. Has been in the Swedish language since 1652.
Laktosfri Fil	lactose-free fil	3.5%	fil culture	Valio		Filmjölk made from 3.5% milkfat and treated with lactase enzyme. Comes unflavoured only.

Probiotic Filmjölk Variants

Name	Literal translation	Milkfat content	Fermentation culture	Produced by	Year introduced	Description
A-fil		0.5%, 2.7%, 3%	fil culture + *Lactobacillus acidophilus*	Arla Foods, Falköpings Mejeri, Gefleortens Mejeri, Milko, Skånemejerier, Wapnö AB	1984 (Arla)	Filmjölk with *Lactobacillus acidophilus*, a commonly used probiotic bacteria. Comes unflavoured and flavoured. Also comes in a low-lactose variant that has been treated with lactase enzyme.
Cultura aktiv fil	Cultura active fil	0.1%	fil culture + *Lactobacillus casei* F19	Arla Foods	2004	Filmjölk with *Lactobacillus casei* F19, a patented probiotic bacteria. Comes unflavoured only.
Kefir		3%	*Lactococcus lactis* subsp. *lactis*, *Lactococcus lactis* subsp. *cremoris*, *Lactobacillus brevis*, *Leuconostoc mesenteroides* subsp. *cremoris*, *Candida kefyr*	Arla Foods	1977	Filmjölk variant based on kefir, a probiotic food; only contains a small subset of microorganisms found in kefir grains. Originated in Caucasus. Comes unflavoured.
Onaka	stomach (Japanese)	1.5%	fil culture + *Bifidobacterium lactis*	Arla Foods	1990	Filmjölk with Bifidobacterium lactis, a probiotic bacteria popular in Japan. Comes unflavoured and flavoured.
Philura		1.5%, 2.6%	*Lactobacillus acidophilus*, *Bifidobacterium lactis*, *Lactobacillus casei*	Milko	2003	Tastes somewhere between regular filmjölk and yogurt. Contains probiotic bacteria that is normally found in the digestive system. Comes unflavoured and flavoured.
Verum hälsofil	Verum health fil	0.5%, 4%	*Lactococcus lactis* L1A	Norrmejerier	1990	Filmjölk that contains at least 10×10^9 *Lactococcus lactis* L1A bacteria per deciliter. Comes unflavoured and flavoured. *Lactococcus lactis* L1A is a patented strain of probiotic bacteria that originated from a culture of långfil from a farm in Västerbotten. In 1998 Verum hälsofil was approved as a natural medical product (*naturläkemedel*) by the Swedish national regulatory agency Medical Products Agency (*Läkemedelsverket*). It has been shown to have a positive effect on the immune and digestive system.

Öresundsfil	Öresund fil	0.9%, 1%	fil culture + *Lactobacillus acidophilus* and *Bifido-bacterium*	Skånemejerier	2000	Filmjölk with *Lactobacillus acidophilus* and *Bifidobacterium*, probiotic bacteria. Comes unflavoured and flavoured.
ProViva Naturell Filmjölk	ProViva unflavoured filmjölk	1%	fil cuture + *Lactobacillus plantarum* 299v	Skånemejerier	1994	Filmjölk that contains at least 50 x 10^6 Lp 299v per milliliter. Comes unflavoured. Lp 299v, a patented probiotic bacteria, has been shown to decrease the symptoms of colon irritation and stressed digestive system in people who consumed ProViva.

Homemade Filmjölk

To make filmjölk, a small amount of bacteria from an active batch of filmjölk is normally transferred to pasteurised milk and then left one to two days to ferment at room temperature or in a cool cellar. The fil culture is needed when using pasteurised milk because the bacteria occurring naturally in milk are killed during the pasteurization process.

A variant of filmjölk called *tätmjölk*, *filtäte*, *täte* or *långmjölk* is made by rubbing the inside of a container with leaves of certain plants: sundew (*Drosera*, Swedish: *sileshår*) or butterwort (*Pinguicula*, Swedish: *tätört*). Lukewarm milk is added to the container and left to ferment for one to two days. More *tätmjölk* can then be made by adding completed *tätmjölk* to milk. In *Flora Lapponica* (1737), Carl von Linné described a recipe for *tätmjölk* and wrote that any species of butterwort could be used to make *tätmjölk*.

Sundew and butterwort are carnivorous plants that have enzymes that degrade proteins, which make the milk thick. How butterwort influences the production of tätmjölk is not completely understood – lactic acid bacteria have not been isolated during analyses of butterwort.

Kumis

Kumis is a fermented dairy product traditionally made from mare's milk. The drink remains important to the peoples of the Central Asian steppes, of Huno-Bulgar, Turkic and Mongol origin: Kazakhs, Bashkirs, Kalmyks, Kyrgyz, Mongols, and Yakuts.

Kumis is a dairy product similar to *kefir*, but is produced from a liquid starter culture, in contrast to the solid *kefir* "grains". Because mare's milk contains more sugars than cow's or goat's milk, when fermented, *kumis* has a higher, though still mild, alcohol content compared to *kefir*.

Even in the areas of the world where *kumis* is popular today, mare's milk remains a very limited commodity. Industrial-scale production, therefore, generally uses cow's milk, which is richer in fat and protein, but lower in lactose than the milk from a horse. Before fermentation, the cow's milk is fortified in one of several ways. Sucrose may be added to allow a comparable fermentation. Another technique adds modified whey to better approximate the composition of mare's milk.

Terminology and Etymology

Mongolian airag (fermented horse milk)

Kumis is also transliterated *kumiss, kumiz, koumiss, kymys, kymyz, kumisz, kymyz,* or *qymyz*. The Russian word, comes from the Turkic word *qımız*. Kurmann derives the word from the name of the Kumyks, one of many Turkic peoples, although this appears to be a purely speculative claim. Clauson notes that *kımız* is found throughout the Turkic language family, and cites the 11th-century appearance of the word in *Dīwān Lughāt al-Turk* written by Kaşgarlı Mahmud in the Karakhanid language.

In Mongolia, the drink is called *airag* or, in some areas, *tsegee*. William of Rubruck in his travels calls the drink *cosmos* and describes its preparation among the Mongols.

Production of Mare's Milk

A 1982 source reported 230,000 horses were kept in the Soviet Union specifically for producing milk to make into *kumis*. Rinchingiin Indra, writing about Mongolian dairying, says "it takes considerable skill to milk a mare" and describes the technique: the milker kneels on one knee, with a pail propped on the other, steadied by a string tied to an arm. One arm is wrapped behind the mare's rear leg and the other in front. A foal starts the milk flow and is pulled away by another person, but left touching the mare's side during the entire process.

A mare being milked in Suusamyr valley, Kyrgyzstan

In Mongolia, the milking season for horses traditionally runs between mid-June and early October. During one season, a mare produces approximately 1,000 to 1,200 litres of milk, of which about half is left to the foals.

Nutritional Properties of Mare's Milk

During fermentation, the lactose in mare's milk is converted into lactic acid, ethanol and carbon dioxide, and the milk becomes an accessible source of nutrition for people who are lactose intolerant.

Before fermentation, mare's milk has almost 40% more lactose than cow's milk. According to one modern source, "unfermented mare's milk is generally not drunk", because it is a strong laxative. Varro's *On Agriculture*, from the 1st century BC, also mentions this: "as a laxative the best is mare's milk, then donkey's milk, cow's milk, and finally goat's milk…"; drinking six ounces (190 ml) a day would be enough to give a lactose-intolerant person severe intestinal symptoms.

Production of Kumis

Kumis is made by fermenting raw unpasteurized mare's milk over the course of hours or days, often while stirring or churning. (The physical agitation has similarities to making butter). During the fermentation, lactobacilli bacteria acidify the milk, and yeasts turn it into a carbonated and mildly alcoholic drink.

Traditionally, this fermentation took place in horse-hide containers, which might be left on the top of a *yurt* and turned over on occasion, or strapped to a saddle and joggled around over the course of a day's riding. Today, a wooden vat or plastic barrel may be used in place of the leather container.

Other accounts from some cities in northern or western China have it that the skin, partially filled with mares' milk, is hung at the door of each home during the season for making such beverages, and passersby, who are familiar with the practice, give each such skin a good punch as they walk by, agitating the contents so they would turn into *kumis* rather than coagulate and spoil.

In modern controlled production, the initial fermentation takes two to five hours at a temperature of around 27 °C (81 °F); this may be followed by a cooler aging period. The finished product contains between 0.7 and 2.5% alcohol.

Kumis itself has a very low level of alcohol, comparable to small beer, the common drink of medieval Europe that also avoided the consumption of potentially contaminated water. *Kumis* can, however, be strengthened through freeze distillation, a technique Central Asian nomads are reported to have employed. It can also be distilled into the spirit known as *araka* or *arkhi*.

History

Archaeological investigations of the Botai culture of ancient Kazakhstan have revealed traces of milk in bowls from the site of Botai, suggesting the domestication of the animal. No specific evidence for its fermentation has yet been found, but considering the location of the Botai culture and the nutritional properties of mare's milk, the possibility is high.

Kumis is an ancient beverage. Herodotus, in his 5th-century BC *Histories*, describes the Scythians processing of mare's milk:

Now the Scythians blind all their slaves, to use them in preparing their milk. The plan they follow is to thrust tubes made of bone, not unlike our musical pipes, up the vulva of the mare, and then to blow into the tubes with their mouths, some milking while the others blow. They say that they do this because when the veins of the animal are full of air, the udder is forced down. The milk thus obtained is poured into deep wooden casks, about which the blind slaves are placed, and then the milk is stirred round. That which rises to the top is drawn off, and considered the best part; the under portion is of less account.

This is widely believed to be the first description of ancient kumis-making. Apart from the idiosyncratic method of mare-milking, it matches up well enough with later accounts, such as this one given by 13th-century traveller William of Rubruck:

This *cosmos*, which is mare's milk, is made in this wise. [...] When they have got together a great quantity of milk, which is as sweet as cow's as long as it is fresh, they pour it into a big skin or bottle, and they set to churning it with a stick [...] and when they have beaten it sharply it begins to boil up like new wine and to sour or ferment, and they continue to churn it until they have extracted the butter. Then they taste it, and when it is mildly pungent, they drink it. It is pungent on the tongue like rapé wine when drunk, and when a man has finished drinking, it leaves a taste of milk of almonds on the tongue, and it makes the inner man most joyful and also intoxicates weak heads, and greatly provokes urine.

Rubruk also mentions that the Mongols prized a particular variety of black kumiss called *caracosmos,* which was made specifically from the milk of black mares.

Health

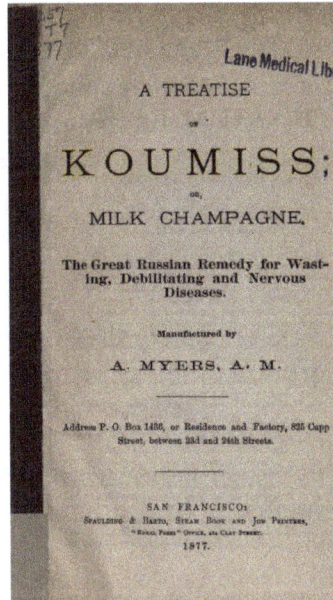

In the West, *kumis* has been touted for its health benefits, as in this 1877 book also naming it "Milk Champagne".

Toward the end of the 19th century, *kumis* had a strong enough reputation as a cure-all to support a small industry of "kumis cure" resorts, mostly in south-eastern Russia, where patients were "furnished with suitable light and varied amusement" during their treatment, which consisted of drinking large quantities of *kumis*. W. Gilman Thompson's 1906 *Practical Dietetics* reported *kumis* has been cited as beneficial for a range of chronic diseases, including tuberculosis, bronchitis, catarrh, and anemia. Gilman also said a large part of the credit for the successes of the "kumis cure" is due not to the beverage, but to favorable summer climates at the resorts. Among notables to try the cure were writers Leo Tolstoy and Anton Chekhov. Chekhov, long-suffering from tuberculosis, checked into a "kumis cure" resort in 1901. Drinking four bottles a day for two weeks, he gained 12 pounds, but no cure.

Consumption

Strictly speaking, *kumis* is in its own category of alcoholic drinks because it is made neither from fruit nor from grain. Technically, it is closer to wine than to beer because the fermentation occurs directly from sugars, as in wine (usually from fruit), as opposed to from starches (usually from grain) converted to sugars by mashing, as in beer. But in terms of experience and traditional manner of consumption, it is much more comparable to beer. It is even milder in alcoholic content than beer and is usually consumed cold. It is arguably the region's beer equivalent.

Kumis is very light in body compared to most dairy drinks. It has a unique, slightly sour flavor with a bite from the mild alcoholic content. The exact flavor is greatly variable between different producers.

As indicated above, *kumis* is usually served cold or chilled. Traditionally it is sipped out of small, handle-less, bowl-shaped cups or saucers, called *piyala*. The serving of it is an essential part of Kyrgyz hospitality on the *jayloo* or high pasture, where they keep their herds of animals (horse, cattle, and sheep) during the summer phase of transhumance.

Cultural Role

The capital of Kyrgyzstan, Bishkek, is named after the paddle used to churn the fermenting milk, showing the importance of the drink in the national culture.

The famous Russian writer Leo Tolstoy in *A Confession* spoke of running away from his troubled life by drinking *kumis*. The Russian composer Alexander Scriabin was recommended a kumis diet and "water cure" by his doctor in his twenties, for his nervous condition and right-hand injury.

The popular Japanese soft drink Calpis models its flavor after the taste of *kumis*.

Kefir

Kefir or kephir, alternatively milk kefir, or búlgaros, is a fermented milk drink made with kefir "grains" (a yeast/bacterial fermentation starter) and has its origins in the north Caucasus Mountains. It is prepared by inoculating cow, goat, or sheep milk with kefir grains. Traditional kefir was made in skin bags that were hung near a doorway; the bag would be knocked by anyone passing through the doorway to help keep the milk and kefir grains well mixed.

Etymology

The word *kefir*, known in Russian since at least 1884, is probably of North Caucasian origin, although some sources see a connection to Turkic *köpür* (foam). Kefir has become the most commonly used term, but other names are found in different geographic regions.

Overview

Kefir grains, a symbiotic culture of bacteria and yeasts

Traditional kefir is fermented at ambient temperatures, generally overnight. Fermentation of the lactose yields a sour, carbonated, slightly alcoholic beverage, with a consistency and taste similar to thin yogurt.

The kefir grains initiating the fermentation are a combination of lactic acid bacteria and yeasts in a matrix of proteins, lipids, and sugars, and this symbiotic matrix, (or SCOBY) forms "grains" that resemble cauliflower. For this reason, a complex and highly variable community of lactic acid bacteria and yeasts can be found in these grains although some predominate; Lactobacillus species are always present. Even successive batches of kefir may differ due to factors such as the kefir grains rising out of the milk while fermenting, or curds forming around the grains, as well as room temperature.

Kefir grains contain a water-soluble polysaccharide known as kefiran, which imparts a creamy texture and feeling in the mouth. The grains range in color from white (the acceptable color of healthy grains), to yellow; the latter is the outcome of leaving the grains in the same milk during fermentation for longer than the optimal 24-hour period, and continually doing so over many batches. Grains may grow to the size of walnuts, and in some cases larger.

The composition of kefir depends greatly on the type of milk that was fermented, including the concentration of vitamin B12.

During fermentation, changes in composition of nutrients and other ingredients occur. Lactose, the sugar present in milk, is broken down mostly to lactic acid (25%) by the lactic acid bacteria, which results in acidification of the product. Propionibacteria further break down some of the lactic acid into propionic acid (these bacteria also carry out the same fermentation in Swiss cheese). A portion of lactose is converted to Kefiran, which is indigestible by gastric digestion. Other substances that contribute to the flavor of kefir are pyruvic acid, acetic acid, diacetyl and acetoin (both of which contribute a "buttery" flavor), citric acid, acetaldehyde and amino acids resulting from protein breakdown.

Kefir preparation

The slow-acting yeasts, late in the fermentation process, break lactose down into eth-anol and carbon dioxide. Depending on the process, ethanol concentration can be as high as 1–2% (achieved by small-scale dairies early in the 20th century), with the kefir having a bubbly appearance and carbonated taste. This makes kefir different from yo-gurt and most other sour milk products where only bacteria ferment the lactose into acids. Most modern processes, which use shorter fermentation times, result in much lower ethanol concentrations of 0.2–0.3%.

As a result of the fermentation, very little lactose remains in kefir. People with lactose intolerance are able to tolerate kefir, provided the number of live bacteria present in this beverage consumed is high enough (i.e., fermentation has proceeded for adequate time). It has also been shown that fermented milk products have a slower transit time than milk, which may further improve lactose digestion.

For the preparation of the present factory-produced kefir, the so-called kefir mild, kefir grains are no longer used, but a precise composed mixture of different bacteria and yeast, allowing the flavor to be kept constant.

Variations that thrive in various other liquids exist, and they vary markedly from kefir in both appearance and microbial composition. Water kefir (or *tibicos*) is grown in wa-ter with sugar (sometimes with added dry fruit such as figs, and lemon juice) for a day or more at room temperature.

Nutrition

Nutritional Composition

Kefir products contain nutrients in varying amounts from negligible to significant con-tent, including dietary minerals, vitamins, essential amino acids, and conjugated lin-oleic acid, in amounts similar to unfermented cow, goat or sheep milk. Kefir is com-posed mainly of water and by-products of the fermentation process, including carbon dioxide and ethanol.

Typical of milk, several dietary minerals are found in kefir, such as calcium, iron, phosphorus, magnesium, potassium, sodium, copper, molybdenum, manganese, and zinc in amounts that have not been standardized to a reputable nutrient database. Also similar to milk, kefir contains vitamins in variable amounts, including vitamin A, vitamin B_1 (thiamine), vitamin B_2 (riboflavin), vitamin B_3 (niacin), vitamin B_6 (pyridoxine), vitamin B_9 (folic acid), vitamin B_{12} (cyanocobalamin), vitamin C, vitamin D, and vitamin E. Essential amino acids found in kefir include methionine, cysteine, tryptophan, phenylalanine, tyrosine, leucine, isoleucine, threonine, lysine, and valine, as for any milk product.

Probiotics

Several varieties of probiotic bacteria are found in kefir products such as Lactobacillus acidophilus, Bifidobacterium bifidum, Streptococcus thermophilus, Lactobacillus delbrueckii subsp. bulgaricus, Lactobacillus helveticus, Lactobacillus kefiranofaciens, Lactococcus lactis, and Leuconostoc species. The significance of probiotic content to nutrition or health remains unproven. Lactobacilli in kefir may exist in concentrations varying from approximately 1 million-1 billion colony-forming units per milliliter and are the bacteria responsible for the synthesis of the polysaccharide kefiran.

In addition to bacteria, kefir often contains strains of yeast that can metabolize lactose, such as Kluyveromyces marxianus, Kluyveromyces lactis and Saccharomyces fragilis as well as strains of yeast that do not metabolize lactose, including Saccharomyces cerevisiae, Torulaspora delbrueckii, and Kazachstania unispora; however, the nutritional significance of these strains is unknown.

Research

A 2003 study found that consumption of the polysaccharide kefiran by human adults with lactose intolerance led to a significant decrease in flatulence.

Production

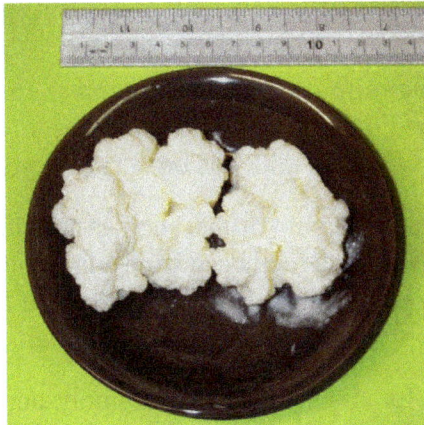

90 grams of kefir grains

Production of traditional kefir requires a starter community of kefir grains which are added to the liquid one wishes to ferment.

The traditional, or artisanal method of making kefir is achieved by directly adding kefir grains (2–10%) to milk in a sealed goatskin leather bag, which is traditionally agitated one or more times a day. Today the leather bag is replaced with a suitable non corrosive container such as a glass jar. It is not filled to capacity, allowing room for some expansion as the carbon dioxide gas produced causes the liquid level to rise. If the container is not light proof it should be stored in the dark to prevent degradation of light sensitive vitamins. After a period of fermentation lasting around 24 hours, ideally at 20–25 °C (68–77 °F), the grains are removed from the liquid by straining using a non-corrosive straining utensil, which can be stainless steel or food grade plastic and reserved as the natural-starter to once again ferment a fresh amount of liquid.

The fermented liquid-kefir which contains live micro-organisms from the grains, may now be consumed as a beverage, used in recipes, or kept aside in a sealed container for many days to undergo a slower secondary fermentation. This process further sours the liquid and through bio-synthesis by certain micro-organisms the content of folic acid and some other B vitamins is increased. Without refrigeration, the shelf life is up to thirty days. The grains will enlarge in the process of kefir production, and eventually split.

The Russian method permits production of kefir on a larger scale, and uses two fermentations. The first step is to prepare the cultures by incubating milk with grains (2–3%), as just described. The grains are then removed by filtration and the resulting liquid mother culture is added to milk (1–3%) which is fermented for 12 to 18 hours.

Kefir can be produced using freeze-dried cultures commonly available as a powder from health food shops. A portion of the resulting kefir can be saved to be used a number of times to propagate further fermentations but ultimately does not form grains, and a fresh culture must be obtained.

Milk Types

Kefir grains will ferment the milk from most mammals, and will continue to grow in such milk. Typical milks used include cow, goat, and sheep, each with varying organoleptic and nutritional qualities. Raw milk has been traditionally used.

Kefir grains will also ferment milk substitutes such as soy milk, rice milk, and coconut milk, as well as other sugary liquids including fruit juice, coconut water, beer wort and ginger beer. However, the kefir grains may cease growing if the medium used does not contain all the growth factors required by the bacteria.

Milk sugar is not essential for the synthesis of the polysaccharide that makes up the grains (kefiran), and studies have shown that rice hydrolysate is a suitable alternative medium. Additionally, it has been shown that kefir grains will reproduce when fermenting soy milk, although they will change in appearance and size due to the differing proteins available to them.

Consumption

Kefir is a popular drink across Eastern and Northern Europe. It was consumed in Russia and Central Asian countries for centuries, but is now becoming popular in Japan, the United States and Europe.

In Chile, where it is known as "yogur de pajaritos" (little birds' yogurt), kefir has been regularly consumed for over a century; it might have been introduced by one of the various waves of migrants from the former Ottoman Empire and migrants from Eastern Europe.

Culinary Uses

Lithuanian kefir-based cold borscht (šaltibarščiai)

As it contains lactobacilli bacteria, kefir can be used to make a sourdough bread. It is also useful as a buttermilk substitute in baking. Kefir is one of the main ingredients in cold borscht in Lithuania and Poland. Other variations of kefir soups, such as kefir-based okroshka, and other foods prepared with kefir are popular across the former Soviet Union and Poland. Kefir may be used in place of milk on cereal, granola or milkshakes.

Possible Origin of Kefir Grains

Kefir grains may be produced by using pasteurized milk inoculated with sheep intestinal flora, followed by culture on the surface of milk.

Other studies indicate small kefir granules may form initially from aggregations of lactobacilli and yeast, followed by a biofilm created by the adherence of additional bacteria and yeasts to the granule exterior.

Chal

Chal, or shubat, is a Turkic (especially Turkmen and Kazakh) beverage of fermented camel milk, sparkling white with a sour flavor, popular in Central Asia — particularly in Kazakhstan and Turkmenistan. In Kazakhstan the drink is known as *shubat*, and is a staple summer food. Due to preparation requirements and perishable nature, chal has proved difficult to export. *Agaran* (fermented cream) is collected from the surface of chal.

Fermented chal is reputed to possess virucidal and virus inhibiting properties not found in fresh camel or cow milk, both in its liquid and lyophilized form — a characteristic which is (reputedly) unaffected by shelf life.

Chal is typically prepared by first souring camel milk in a skin bag or ceramic jar by adding previously soured milk. For 3–4 days, fresh milk is mixed in; the matured chal will consist of one third to one fifth previously soured milk.

Camel milk will not sour for up to 72 hours at temperatures below 10 °C (50 °F). At 30 °C (86 °F) the milk sours in approximately 8 hours (compared to cow's milk, which sours within 3 hours).

A comparison of the composition of camel milk and camel chal:

	Camel milk	"Chal"
acidity	18°D	28°D
fat	4.3%	4.3%
lactose	2.75%	1.32%
non-fat solids	8.2%	6.6%
ash	0.86%	0.75%
ethyl alcohol		1.1%
ascorbic acid	5.6 mg%	4.8 mg%

Dornic acidic degrees are used to describe acidity in milk products, with 1 Dornic degree (1°D) is equal to 0.1g of lactic acid per liter. The chal contained Lactobacilli lactic; streptococci and yeast.

Chal may be cultured with lactobacillus casei, streptococcus thermophilus and lactose-fermenting yeasts incubating in inoculated milk for 8 hours at 25 °C (77 °F), and then subsequently for 16 hours at 20 °C (68 °F). Holder pasteurization does not affect the quality of the milk, but pasteurization at higher temperatures (85 °C/185 °F) for 5 minutes negatively impacts flavour. Chal made from pure cultures of lactobacillus casei, streptococcus thermophilus and species of torula has markedly less not-fat solids and lactose than the milk from which it is made.

Buttermilk

Buttermilk (right) compared to fresh milk (left). The thicker buttermilk leaves a more visible residue on the glass.

Buttermilk refers to a number of dairy drinks. Originally, buttermilk was the liquid left behind after churning butter out of cream. This type of buttermilk is known as *traditional buttermilk*.

The term *buttermilk* also refers to a range of fermented milk drinks, common in warm climates (e.g., the Balkans, the Middle East, Turkey, Afghanistan, Pakistan, Nepal, India, Sri Lanka, Nicaragua and the Southern United States) where unrefrigerated fresh milk sours quickly, as well as in colder climates, such as Scandinavia, Ireland, the Netherlands, Germany, Poland, Slovakia, Slovenia, Croatia and the Czech Republic. This fermented dairy product known as *cultured buttermilk* is produced from cow's milk and has a characteristically sour taste caused by lactic acid bacteria. This variant is made using one of two species of bacteria—either *Lactococcus lactis* or *Lactobacillus bulgaricus*, which creates more tartness.

The tartness of buttermilk is due to acid in the milk. The increased acidity is primarily due to lactic acid produced by lactic acid bacteria while fermenting lactose, the primary sugar in milk. As the bacteria produce lactic acid, the pH of the milk decreases and casein, the primary milk protein, precipitates, causing the curdling or clabbering of milk. This process makes buttermilk thicker than plain milk. While both traditional and cultured buttermilk contain lactic acid, traditional buttermilk tends to be less viscous, whereas cultured buttermilk is more viscous.

Buttermilk can be drunk straight, and it can also be used in cooking. Soda bread is a bread in which the acid in buttermilk reacts with the rising agent, sodium bicarbonate, to produce carbon dioxide which acts as the leavening agent. Buttermilk is also used in marination, especially of chicken and pork, whereby the lactic acid helps to tenderize, retain moisture, and allows added flavors to permeate throughout the meat.

Traditional Buttermilk

Originally, buttermilk referred to the liquid left over from churning butter from cultured or fermented cream. Traditionally, before cream could be skimmed from whole milk, the milk was left to sit for a period of time to allow the cream and milk to separate. During this time, naturally occurring lactic acid-producing bacteria in the milk fermented it. This facilitates the butter churning process, since fat from cream with a lower pH coalesces more readily than that of fresh cream. The acidic environment also helps prevent potentially harmful microorganisms from growing, increasing shelf-life. However, in establishments that used cream separators, the cream was hardly acidic at all.

On the Indian subcontinent, the term "buttermilk" refers to the liquid left over after extracting butter from churned cream. Today, this is called *traditional buttermilk*. Traditional buttermilk is still common in many Indian, Nepalese, and Pakistani households, but rarely found in Western countries. In Southern India and in the states of Punjab, Gujarat and Rajasthan, buttermilk topped with sugar, salt, cumin, asafoetida, or curry leaves is a common accompaniment in every meal.

Cultured Buttermilk

Commercially available cultured buttermilk is milk that has been pasteurized and homogenized (with 1% or 2% fat), and then inoculated with a culture of Lactococcus lactis (formerly known as Streptococcus lactis) plus Leuconostoc citrovorum to simulate the naturally occurring bacteria in the old-fashioned product. Some dairies add colored flecks of butter to cultured buttermilk to simulate residual flecks of butter that can be left over from the churning process of traditional buttermilk.

Condensed buttermilk and dried buttermilk have increased in importance in the food industry. Buttermilk solids are used in ice cream manufacturing, as well as being added to pancake mixes. Adding specific strains of bacteria to pasteurized milk allows more consistent production.

In the early 1900s, cultured buttermilk was labeled *artificial buttermilk*, to differentiate it from traditional buttermilk, which was known as *natural* or *ordinary buttermilk*.

Acidified Buttermilk

Acidified buttermilk is a related product made by adding a food-grade acid (such as lemon juice) to milk. It can be produced by mixing 1 tablespoon of vinegar or lemon juice per 1 cup of milk and letting it sit until it curdles, about 10 minutes. Any level of fat content for the milk ingredient may be used, but whole milk is usually used for baking. In the process which is used to produce paneer, such acidification is done in the presence of heat.

Powdered Buttermilk

Like powdered milk, buttermilk is available in a dried powder form. This stores well at room temperature and is usually used in baked goods.

Nutrition

One cup (237 mL) of whole milk contains 157 calories and 8.9 grams of fat whereas one cup of buttermilk contains 99 calories and 2.2 grams of fat. Buttermilk contains vitamins, potassium, calcium, and traces of phosphorus.

Chaas

A glass full of Chaach

Chaas, also pronounced 'Chaach' and known by several other names in other Indian languages, is a yogurt-based drink popular across India. People in India have a tendency to translate Chaas into English as "buttermilk," but this would be a misnomer, and would convey an incorrect meaning to people whose mother-tongue is English.

Various Names

Chaas is the name by which this beverage is known in the Gujarati and Urdu languages, and in some regions of Hindi-speaking north India. It is known as *Mattha* in other parts of Hindi-speaking north India, as *Mor* in Tamil, as *Mooru* in Malayalam, as *majjige* in Kannada and Telugu, as *taak* or *tak* in Marathi and as *ghol* in Bengali.

Preparation and Variations

Chaas is made by churning yogurt (curds/dahi) and cold water together in a pot, using a hand-held instrument called *madhani* (whipper). This can be consumed plain or seasoned with a variety of spices. Chaas can be made from fresh yogurt, and the natural flavour of such Chaas is mildly sweet. This type of Chaas is very close to Lassi, with two major differences: Chaas is more dilute (with water) than lassi and unlike lassi, Chaas does not have added sugar.

Although Chaas can be made from fresh yogurt (curds/dahi), it is more commonly made at home from yogurt that is a few days old and has become sour due to age. Indeed, one of the purposes for making Chaas at home is to usefully finish off old yogurt that is lying in the fridge for long. Such Chaas has a tangy, slightly sour taste which is considered delicious. A pinch of salt is usually added to it for further enhancement of taste, and other seasonings can be added also, as described below.

A third variation of Chaas is obtained by adding actual buttermilk (water left over after churning butter) into the Chaas. This gives a slightly sour-bitter taste to the final product, and it is necessary to add seasonings to mask these flavours. Chaas made using buttermilk is very healthy but the taste is not relished by all. However, if proper seasonings and spices are used, it can be delicious. This type of Chaas is more unusual and rare compared to the other types, because it is available only when butter is churned at home.

Seasoning and Flavours

Chaas can be consumed plain, but a little salt is usually added. This is the most common seasoning for Chaas. Numerous other seasonings and spices can be added to salted Chaas, either singly or in combination with each other. These spices are usually roasted in a wok, using a spoonful of cooking oil, before being added to the Chaas. The spices which can be added thus are: Coarsely ground and roasted cumin seeds, curry leaves, asafoetida, grated ginger, very finely diced green chillies and Mustard seeds.

Sugar can also be added to Chaas, but if sugar is added, than neither salt nor spice can be used. Adding sugar to Chaas makes it very similar to lassi, the main difference being that Chaas is more dilute (with water) than lassi. Lassi is more popular in Punjab and certain regions of north India, while Chaas (known by various named) is popular in all other parts of the country.

Vendors have come up with several proprietary products and standardized flavours of Chaas which are produced on an industrial scale and sold as bottled drinks. The best-seller among such brands is Amul's *Masala Chass,* which has standardized several traditional flavours for the mass bottled-drink market. Other popular modern flavours which are only available as bottled drinks and cannot be made at home or in restaurants include rose-flavoured Chaas ("Chaas Gulabi") and Chaas flavoured with mint ("Mint Chaas"). Both these flavours are of the added-sugar variety and are different from flavoured lassi only for being more dilute and less expensive.

Consumption and Benefits

In India, the consumption of Chaas has cultural resonances and associations which are not found in the context of other beverages like tea, coffee or lassi. Chaas is associated with two major benefits: cooling and improved digestion. Both of these properties are improved by the addition of spices.

The cooling properties of Chaas, which helps people to rehydrate and beat the heat of the Indian summer, is greatly valued traditionally. Chaas is considered as a cool drink which keeps the body temperature down in summer. An earthen pot is used to prepare chaas and store it for a few hours before consumption. The use of earthen pot makes the chaas cool even in summer. In the extremely hot desert areas of Gujarat and Rajasthan people consume chaas with salt after getting exposed in the sun because this cools the body and aids in rehydration. In the summer months, crushed ice is often added to the Chaas.

Chaas is consumed all year round. It is usually taken immediately after meals, but is also consumed on its own as a beverage. If spices (especially jeera cumin) are added to the Chaas, it can improve digestion.

Lassi

Lassi is a popular traditional yogurt-based drink from the Indian Subcontinent. Lassi is a blend of yogurt, water, spices and sometimes fruit. *Traditional* lassi (a.k.a., "salted lassi", or simply "lassi") is a savoury drink, sometimes flavoured with ground and roasted cumin. *Sweet* lassi, however, contains sugar or fruits, instead of spices. Salted mint lassi is highly favoured in Bangladesh.

In Dharmic religions, yogurt sweetened with honey is used while performing religious rituals. Less common is lassi served with milk and topped with a thin layer of clotted cream. Lassis are enjoyed chilled as a hot-weather refreshment, mostly taken with lunch. With a little turmeric powder mixed in, it is also used as a folk remedy for gastroenteritis. In Pakistan, salted lassi is often served with almost all kinds of meals, and is mostly made at home by simply whisking salt in yogurt and water. It is also sold at most dairy shops selling yogurt and milk, and both the salty and sweet variety are available.

Variations

Traditional Mild Sweet (or Salty) Lassi

Traditional mild sweet (or salty) form of lassi is more common in North India and Punjab, Pakistan. It is prepared by blending yogurt with water and adding sugar and other spices to taste. Salt can be substituted in place of sugar. The resulting beverage is known as salted lassi. This is similar to doogh.

Salt lassi.

Mint sweet lassi in Chandigarh, India.

Sweet Lassi

Sweet lassi is a form of lassi flavoured with sugar, rosewater and/or lemon, strawberry or other fruit juices. Saffron lassis, which are particularly rich, are a specialty of Rajasthan and Gujarat in India and Sindh in Pakistan. *Makkhaniya lassi* is simply lassi with lumps of butter in it (*makkhan* is the Gujarati, Hindi, Sindhi and Punjabi word for butter). It is usually creamy like a milkshake.

Mango Lassi

Mango lassi is gaining popularity worldwide. It is made from yogurt, water and mango pulp. It may be made with or without additional sugar. It is widely available in UK, Malaysia, Singapore, the United States, and in many other parts of the world. In various parts of Canada, mango lassi is a cold drink consisting of sweetened kesar mango pulp mixed with yogurt, cream, or ice cream. It is served in a tall glass with a straw, often with ground pistachio nuts sprinkled on top.

Bhang Lassi

Bhang lassi is a special, narcotic lassi that contains bhang, a liquid derivative of cannabis, which has effects similar to other eaten forms of cannabis. It is legal in many parts of India and mainly sold during Holi, when pakoras containing bhang are also sometimes eaten. Uttar Pradesh is known to have licensed bhang shops, and in many places one can buy bhang products and drink bhang lassis.

Chaas

Chaas or *chaach* is a salted drink like lassi, difference being that *chaas* contains more water than lassi and has the butterfat removed, so its consistency is not as thick as lassi. Salt and *jeera* (cumin seeds) is usually added for taste and sometimes even fresh coriander. Fresh ground ginger and green chillies may also be added as seasoning. *Chaas* is popular in the Indian states of Gujarat and Rajasthan, where it is a common accompaniment at mealtime. It aids digestion and is an excellent coolant in the Indian and Pakistani summers . It is called *majjige* in Kannada, *taak* in Marathi, *majjiga* in Telugu, *moru* in Tamil and Malayalam, *mahi* in the Madheshi languages, and *ghol* in Bengali.

Cultural References

A 2008 print and television ad campaign for HSBC, written by Jeffree Benet of JWT Hong Kong, tells a tale of a Polish washing machine manufacturer's representative sent to India to discover why their sales are so high there. On arriving, the representative investigates a lassi parlor, where he is warmly welcomed, and finds several washing machines being used to mix it. The owner tells him he is able to "make ten times as much lassi as I used to!"

On his *No Reservations* television program, celebrity chef Anthony Bourdain visited a "Government Authorised" Bhang Shop in Jaisalmer Fort, Rajasthan. The proprietor offered him three varieties of bhang lassi: "normally strong, super duper strong, and full power 24 hour, no toilet, no shower."

In 2013, Kshitij, the annual techno-management fest of IIT Kharagpur, had launched a campaign to name the next version of the mobile operating system Android, Lassi.

Yakult

Yakult is a probiotic dairy product made by fermenting a mixture of skimmed milk with a special strain of the bacterium *Lactobacillus casei* Shirota. It was created by Japanese scientist Minoru Shirota, who graduated from the Medical School of Kyoto University in 1930. In 1935, he started manufacturing and selling Yakult. Official claims state that the name is derived from *jahurto*, an Esperanto word for "yogurt". Since then, Yakult has also introduced a line of beverages for the Japanese market that contain *Bifidobacterium breve* bacteria, and has also used its lactobacilli research to develop cosmetics. More recently, the Yakult Honsha played a major role in developing the chemotherapy drug irinotecan (Camptosar, CPT-11).

After its introduction in Japan and Taiwan, Yakult was first sold in Brazil in 1966, due to the large number of Japanese immigrants in the country, before it was marketed elsewhere. Today, Yakult is sold in 31 countries, although its bacteria cultures are provided from a mother strain from Japan regardless of production location.

Yakult is marketed in different sizes. In Australia and New Zealand, Europe, India, Indonesia, and Vietnam Yakult comes in 65mL bottles. In America (including Mexico, one of Yakult's largest selling markets), Japan, Philippines, Thailand, Malaysia and South Korea, 80 ml bottles are available. It is also available in Singapore, Hong Kong, Taiwan and China where Yakult comes in 100 ml bottles.

The product is made by Yakult Honsha Co., Ltd. (株式会社ヤクルト本社 *Kabushi-ki-gaisha Yakuruto Honsha?*) (TYO: 2267), where Groupe Danone has a 20% share. The company also owns one of Japan's major baseball franchises, the Tokyo Yakult Swallows. It has been also one of the partner companies of the FINA World Aquatics Championships since 2005.

Nutritional Value

Standard Yakult (excludes variations such as in Yakult Light) contains:

- Sugar (sucrose, dextrose)
- Skimmed milk powder
- Natural flavours
- Live *Lactobacillus casei* Shirota strain, 6.5 billion per 65 ml bottle (concentration of 10^8 CFU/mL)
- Water

Standard Yakult contains 14g of sugar for every 100g, but comes in 65 ml bottles. This concentration is higher than the level defined as "HIGH" by the UK Food Standards Agency (described for concentrations of sugar above 15g per 100g). As a comparison Coca-Cola and orange juice are around 10g of sugar per 100g, but with a serving size usually greater than 250 ml, while Yakult is served in much smaller doses. Based on the existence of many products like Yakult in the world market and the content of milk protein (1.4 g per 100 ml), it was classified as a fermented milk named "Fermented Milk

Drinks" of the Codex Standard. Through nutrient profiling guidelines, current health claim regulation in European Union may forbid the use of health claim on food products that are nutritionally unbalanced, but dairy products and probiotic drinks are likely to be considered as favourable carrier because their health benefits outweigh the fact they might be high in one of the designated 'unhealthy' ingredients.

Scientific Basis

According to the manufacturer's website, the benefits of Yakult consumption are supported by an array of scientific studies. Those could range from maintenance of gut flora, "modulation" of the immune system, regulation of bowel habits and constipation and finally effects on some gastrointestinal infections. Although the number of scientific papers is large, most of them are based on *in vitro* and *in vivo* experiments, with some human clinical trials done on cohorts and with daily consumption of 40–100 billions of probiotic *L. casei* Shirota, far above the single bottle concentration of approximately 6.5 billion. In 2010, an EFSA panel concluded that a cause and effect relationship has not been established between the consumption of Lactobacillus casei strain Shirota and maintenance of the upper respiratory tract defence against pathogens by maintaining immune defences. In September 2014, University College London published the results of testing using stomach fluids from pigs. The tests found that Yakult had sufficient bacteria, but "fell short" as the bacteria were unable to survive in the stomach. Yakult responded by claiming that independent studies had shown the bacteria do better in the human digestive system.

Production

Yakult opened a factory in Fountain Valley, California in the United States in 2014. Yakult is also manufactured in Australia in Dandenong, Victoria. A manufacturing plant of Yakult is situated at Haryana, India.

In Malaysia, Yakult is manufactured at their factory located at Seremban, Negeri Sembilan.

Yakult opened a factory in Calamba, Laguna in the Philippines due to their popularity.

Varieties

| Yakult Light 65ml available in Australia. | Yakult 300 Light 100ml available in Taiwan. |

Yakult Around the World

| Greater China (Canton district excluded) Yakult. | Yakult 80 ml available in Mexico. |

References

- Zeder, Melinda A. ed. (2006). Documenting Domestication: New Genetic and Archaeological Paradigms. University of California Press. p. .264. ISBN 0-520-24638-1.

- Kurmann, Joseph A.; et al. (1992). Encyclopedia of Fermented Fresh Milk Products. Springer. p. 174. ISBN 0-442-00869-4.

- Steinkraus, Keith H. ed (1995). Handbook of Indigenous Fermented Foods. Marcel Dekker. p. 304. ISBN 0-8247-9352-8.

- Law, B A ed. (1997). Microbiology and Biochemistry of Cheese and Fermented Milk. Springer. p. 120. ISBN 0-7514-0346-6.

- Kowsikowski, F. and Mistry, V. (1997). Cheese and Fermented Milk Foods, 3rd ed, vol. I. F. V. Kowsikowski, Westport, Conn., ISBN 0-9656456-0-6.

- Anatoly Michailovich Khazanov (15 May 1994). Nomads and the outside world (2nd ed.). Univ of Wisconsin Press. p. 49. ISBN 978-0-299-14284-1.

- Sheridan, Paul (2015-05-30). "How to Make Kumis the Scythian Way". Anecdotes from Antiquity. Retrieved 2015-08-27.

- "Nutrition facts for fluid sheep milk, one US cup, 245 ml". Conde Nast, Nutritiondata.com, USDA Nutrient Database, Standard Reference, version 21. 2014. Retrieved 19 November 2014.

- Farnworth, Edward R (4 April 2005). "Kefir-a complex probiotic" (PDF). Food Science and Technology Bulletin: Functional Foods. 2 (1): 1–17. doi:10.1616/1476-2137.13938. Retrieved 20 December 2014.

Permissions

All chapters in this book are published with permission under the Creative Commons Attribution Share Alike License or equivalent. Every chapter published in this book has been scrutinized by our experts. Their significance has been extensively debated. The topics covered herein carry significant information for a comprehensive understanding. They may even be implemented as practical applications or may be referred to as a beginning point for further studies.

We would like to thank the editorial team for lending their expertise to make the book truly unique. They have played a crucial role in the development of this book. Without their invaluable contributions this book wouldn't have been possible. They have made vital efforts to compile up to date information on the varied aspects of this subject to make this book a valuable addition to the collection of many professionals and students.

This book was conceptualized with the vision of imparting up-to-date and integrated information in this field. To ensure the same, a matchless editorial board was set up. Every individual on the board went through rigorous rounds of assessment to prove their worth. After which they invested a large part of their time researching and compiling the most relevant data for our readers.

The editorial board has been involved in producing this book since its inception. They have spent rigorous hours researching and exploring the diverse topics which have resulted in the successful publishing of this book. They have passed on their knowledge of decades through this book. To expedite this challenging task, the publisher supported the team at every step. A small team of assistant editors was also appointed to further simplify the editing procedure and attain best results for the readers.

Apart from the editorial board, the designing team has also invested a significant amount of their time in understanding the subject and creating the most relevant covers. They scrutinized every image to scout for the most suitable representation of the subject and create an appropriate cover for the book.

The publishing team has been an ardent support to the editorial, designing and production team. Their endless efforts to recruit the best for this project, has resulted in the accomplishment of this book. They are a veteran in the field of academics and their pool of knowledge is as vast as their experience in printing. Their expertise and guidance has proved useful at every step. Their uncompromising quality standards have made this book an exceptional effort. Their encouragement from time to time has been an inspiration for everyone.

The publisher and the editorial board hope that this book will prove to be a valuable piece of knowledge for students, practitioners and scholars across the globe.

Index

www.ingramcontent.com/pod-product-compliance
Lightning Source LLC
Chambersburg PA
CBHW061933190326
41458CB00009B/2731

* 9 7 8 1 6 3 5 4 9 0 8 3 1 *